RF-Frontend Design for Process-Variation-Tolerant Receivers

ANALOG CIRCUITS AND SIGNAL PROCESSING SERIES

Series Editors:
Mohammed Ismail, *The Ohio State University*
Mohamad Sawan, *École Polytechnique de Montréal*

For further volumes:
http://www.springer.com/series/7381

Pooyan Sakian • Reza Mahmoudi
Arthur van Roermund

RF-Frontend Design for Process-Variation-Tolerant Receivers

 Springer

Pooyan Sakian
Eindhoven University of Technology
Eindhoven, The Netherlands

Reza Mahmoudi
Eindhoven University of Technology
Eindhoven, The Netherlands

Arthur van Roermund
Eindhoven University of Technology
Eindhoven, The Netherlands

ISBN 978-1-4939-0223-1   ISBN 978-1-4614-2122-1 (eBook)
DOI 10.1007/978-1-4614-2122-1
Springer New York Dordrecht Heidelberg London

Printed on acid-free paper

Springer is part of Springer Science+Business Media (www.springer.com)

Contents

Chapter 1
Introduction

Data exchange and data processing are two vital elements in our age of information which is distinguished by the possibility of free and instant access to knowledge for public. Since the experimental verification of the existence of electromagnetic waves by Heinrich Hertz in 1887, which was literally the first implementation of transmission and reception of radio electromagnetic waves, and achieving wireless transatlantic communication by Marconi in 1901, the wireless applications have observed a tremendous growth, as a widespread means of information exchange.

The desire for higher data rates and thus higher bandwidth, as the most valuable resource in communication systems, has encouraged wireless system designers to exploit unlicensed bands, for instance around 2.4, 5.8, and 60 GHz. The license-free bands around 2.4 and 5.8 GHz are already congested with different applications and also are prone to different interfering communication signals in the adjacent channels. However, the 7 GHz unlicensed band around 60 GHz provides a relatively much higher bandwidth, making it a tempting candidate for wireless solutions to different high data rate applications.

The growth of wireless applications is accompanied with and supported by the remarkable progress in information processing technologies. The huge and ultra-expensive supercomputers of 1970s are now far surpassed in performance by publicly available small notebooks. While many sciences and disciplines have contributed to this evolution, the art of electronics and specifically CMOS transistor circuit design have definitely played an undeniable role. The rapid scaling of CMOS into deep submicron regimes has led to higher performance and higher levels of integration for digital circuits, i.e., higher speed and higher memory capacity in a smaller size. In fact in the past 30 years, device scaling has been the chief support for the integrated circuit industries to maintain the delivery of high performance products at lower costs to the customers [1]. The urge to unify the data transfer and data processing systems on a single chip is one of the main motivations to realize the data transfer (the transceivers) in the same CMOS technology as the digital backend circuitry is implemented with.

P. Sakian et al., *RF-Frontend Design for Process-Variation-Tolerant Receivers*, Analog Circuits and Signal Processing, DOI 10.1007/978-1-4614-2122-1_1, © Springer Science+Business Media New York 2012

Random process variations, always been a problem in integrated circuit design, are significantly aggravated as a result of technology scaling beyond 90 nm [2]. The impact has been so severe that it has almost shifted the digital design problem from deterministic to probabilistic [3–5]. In some cases the speed improvement of digital circuits resulting from transistor scaling is canceled out at the worst case corners just as a consequence of the pronounced impact of process variations [2]. In fact, one of the greatest challenges caused by the nanometer scaling is the raise in the intra-die variability of the threshold voltage and leakage current as they are influenced by the statistical distribution of parameters such as physical gate length and dopant concentration. The variability makes the overdrive voltage unpredictable even for neighboring identically-sized transistors, rendering the digital gate delay a stochastic variable [6]. As the process variations are increased with scaling in the nanometer regime, designers have to develop new methods to implement reliable electronic systems with uncertain components, because the traditional solution of worst-case design turns out to be no more efficient, as the worst case and nominal performance diverge increasingly. The adverse impact of such conservative worst-case design approach is well illustrated by the extent to which the microprocessors can be overclocked. For instance, a microprocessor with nominal clock frequency of 2.2 GHz can be overclocked up to 3.1 GHz [7]. While the process variations can be so influential in the digital design and are extensively addressed in the literature, their importance in the design of wireless and RF frontend circuits cannot be neglected either [8, 9].

Consequently, in this book the design of some circuits required for 60 GHz wireless receivers is investigated and problems related to process variations are addressed.

We divide the problem into two categories of general and specific.

The general problem originates from the regular challenges of designing at 60 GHz and the specific problem is addressing the shortcomings brought up by process variations.

Designing at 60 GHz requires dealing with multiple challenges which might be irrelevant or negligible at low frequency designs. Performing measurements using on-wafer probing at 60 GHz has its own complexities. The very short wave-length of the signals at mm-wave frequencies makes the measurements very sensitive to the effective length and bending of the interfaces. Especially to perform on-wafer measurements one must pay utmost attention to the rigidity of the interfaces connected to the probes to keep all the connection lengths and orientations constant during the whole period of the measurement and calibration. Also special care must be taken to preserve the position of the probes on the bondpads and impedance standard substrates, since the measurement accuracy can be very much dependent on the positioning and landing of the probes. Another difficulty of mm-wave measurements arises from the overwhelming cost of equipment needed for instrumentation. For instance, s-parameter measurement of a differential two-port mm-wave circuit would require a very expensive four-port network analyzer.

One of the most important challenges of 60 GHz circuit design occurs in the transition between schematic and layout. Modeling the performance of circuits after doing the layout and taking into account the parasitic effects resulting from the layout are two issues that are more important and influential at high frequency design. The pronounced impact of parasitics at such high frequencies makes it more difficult to obtain the desired level of performance from the circuits. In addition, the necessity of accurate modeling of the parasitic effects brings about another design complexity. In fact, these complexities lead to the necessity of an iterative shift of the design focus from the schematic to the layout and vice versa, rendering the design a more time consuming process.

The skin effect, which is intensified at these frequencies, raises other difficulties such as increased resistance of the metal layers and additional loss in the signal lines and passive circuits. In addition, the substrate losses are exacerbated as a result of increased electric coupling between metal lines and the substrate.

Working closer to the cutoff frequency (f_T) of the transistors, especially in technologies with relatively low f_T, makes it more difficult to achieve the desired gain from the amplifiers. In addition, the low quality factor of the varactors at millimeter-wave regime is another source of complication, particularly for designing voltage-controlled oscillators (VCO) and reconfigurable radio-frequency (RF) circuits.

The electromagnetic modeling of complex structures including the coupling impact of adjacent components is another issue which is sometimes impractical with the currently available simulation software, as they may require immense computational power.

Therefore, the question facing the designers is whether the currently available software and tools are computationally capable of including all the layout impacts in their prediction of the performance of the circuits and how such predictions can be accurate regarding all the aforementioned limitations and the accentuated impact of layout-level issues.

Process-induced variability engenders performance deteriorations in RF circuits, whereas process-induced mismatches result in problems like augmented second order intermodulation distortion or I-Q mismatch in homodyne receivers. Scaling into the deep submicron regime, mainly in CMOS technologies, accentuates the effect of process spread and mismatch on the fabrication yield [10]. Design for manufacturability requires all manufacturing and process variations to be considered in the design procedure. Designers have been traditionally required to manage these variations by designing for worst-case device characteristics (usually, a three-sigma variation from nominal conditions), which results in excessively conservative designs [11]. Statistical circuit-level methods based on modeling data provided by fabrication foundries, e.g. Monte Carlo, are extensively used to evaluate the effect of process spread and are utilized by simulation tools to design circuits with the desired performance over the specified range of process variation [8]. However, most of these statistical methods are based on random variation of design variables which need long simulation times for large-scale circuits, like a full receiver. Furthermore, as the size and complexity of designs is increased, less insight is obtained from these random statistical methods.

Therefore, the question remains as to whether there is any design guideline with analytical basis and involving less complexity for designing receivers that are robust to process variations. What would be the limitations of such a design guideline? Is it possible to improve the resilience of a receiver to process-induced variability by implementing adaptive building blocks? What is the best location in the receiver for exploiting adaptive components? Is it possible to adapt the performance of the receiver by tuning a small set of parameters to achieve a performance correction scheme suitable for both operation-time calibration and production-line trimming? Due to the limited time and scope of this work we investigate answers to these questions only for the receivers.

Splitting the problem into two categories of, first, 60 GHz circuit design and measurement challenges and, second, process-induced variability, the solution approach is also divided into two parts.

In the first part we have to take care of proper and meticulous modeling of layout effects to improve the predictability of the performance. In addition, the layout should be done in a way to minimize the parasitic capacitances whenever they are undesirable. Appropriate layout measures should be taken to reduce the performance degradation in the post-layout design. For modeling complex structures we have to use electromagnetic simulators or use valid analytical models to calculate the impacts. Obviously, this leads to an iterative design process shifting from the circuit schematic to the layout and vice versa.

In the second part, i.e., process-induced variability, we first find an analytical high-level description of the whole receiver system and the dependency and sensitivity of the overall performance of the receiver to the performance of its building blocks, without engaging with circuit-level details which can complicate the analysis. For this purpose, every building block of the receiver should be accurately expressed with a small set of performance parameters which can be applicable to both the receiver as a whole as well as to its constituting components. Such performance parameters must represent the effect of process variations. Then we should investigate the possibility of minimizing the sensitivity of the overall performance to the variability of this set of parameters.

We should also investigate the feasibility of improvements in the robustness of the receiver by circuit-level methods that provide re-configurability to the RF circuits. Such re-configurability must be realized with few tunable circuit parameters to enable the receiver for either automatic calibration during the operation or performance trimming in the production line during the fabrication.

The aim of this book is to develop design guidelines for receivers with robustness to process variations and realize circuit blocks demanded by such design guidelines while obtaining the required performance level.

As a consequence, the work of this book is limited to the receiver and the problems regarding the transmitter are not addressed. In the system-level studies, the whole RF front-end and the baseband circuitry including the ADC are addressed. Special cases for zero-intermediate-frequency (zero-IF) receivers are derived and used in the rest of the book.

The circuit-level work is limited to zero-IF 60 GHz receiver components. Stand-alone components including low-noise amplifier (LNA), mixer, and VCO are designed and measured. Therefore, circuits which are (solely) suitable for other receiver architectures are not addressed in this book. The zero-IF receivers have some advantages in terms of the fewer count of components, compactness, and capability of integration. The other advantage is their potential for future applications at 60 GHz which may require the downconversion of the whole 7 GHz unlicensed band.

All the circuits are designed in CMOS; some in 65 nm and the rest in 45 nm technology. The reason, aside from the advantages of CMOS technology for integration with the digital backend, is the availability of the technology in the course of the project.

In Chap. 2, a system-level sensitivity analysis is performed on a generic RF receiver. The sensitivities of the overall performance with respect to the block-level gain, noise, and linearity are calculated. Methods are derived for reducing some of these sensitivities. The analysis is applied to a zero-IF three-stage 60 GHz receiver. In addition, a system-level study is performed on a 60 GHz receiver including ADC.

In Chap. 3, mm-wave layout challenges and considerations such as the impact of parasitics, substrate losses, cross-talk issues, and electromagnetic modeling of complex structures are explained. Then, the setups used for measuring the 60 GHz circuits designed in this work are illustrated.

In Chap. 4, several 60 GHz components are presented and designed in standard CMOS technologies with intrinsically high performance without exhibiting re-configurability for post-fabrication performance fine tuning. The design, fabrication, and measurement of a low-noise amplifier (LNA), a zero-IF mixer, and a quadrature voltage-controlled oscillator (VCO) are described.

In Chap. 5, circuit level solutions are presented for coping with the impact of process variations. One of the problems associated with process-variation-induced mismatch is the second order intermodulation distortion (IMD2). A tunable mixer is implemented for correcting the mismatches and minimizing the IMD2. Then, by accumulating the noise and nonlinearity contributions in one stage the overall performance of the receiver is made more sensitive to the noise and nonlinearity of that single stage and less sensitive to the noise and nonlinearity of the other stages. This way the performance degradations resulting from process variations can be compensated mostly by tuning the performance of that single stage. This idea is simulated at circuit level in this chapter.

Chapter 6 brings the conclusion and recommendations.

Chapter 2
System-Level Design for Robustness

The increasing demand for compactness and speed of digital circuits and the necessity of integration of the digital backend electronics with radio frequency frontends, calls for exploiting deep submicron technologies in RF circuit design. However, scaling into the deep submicron regime, mainly in CMOS technologies, accentuates the effect of process spread and mismatch on the fabrication yield [10]. Furthermore, design for manufacturability requires all manufacturing and process variations to be considered in the design procedure. Statistical circuit-level methods based on modeling data provided by fabrication foundries, e.g. Monte Carlo, are extensively used to evaluate the effect of process spread and are utilized by simulation tools to design circuits with the desired performance over the specified range of process variation [8]. However, most of these statistical methods are based on random variation of design variables which need long simulation times for large-scale circuits, like a full receiver. Furthermore, as the size and complexity of designs is increased, less insight is obtained from these random statistical methods.

Recently, system-level design techniques are developed to determine the specifications of individual blocks of a receiver for minimum overall power consumption and for a given noise and nonlinearity performance [12–14]. However, system-level design guidelines in the literature do not sufficiently address the sensitivity of the receiver to process variations. In this chapter, for a given required overall performance, the block-level budgeting is performed in such a way that the effect of process variation on the overall performance is minimized. A system-level sensitivity analysis is performed on a generic receiver which can pinpoint the sensitive building blocks and show how to reduce the overall sensitivity to the performance of individual building blocks. Based on the presented analysis the optimum plan for block-level specifications in terms of sensitivity to variability of components can be determined.

In this chapter, each building block of the receiver is described by three process-sensitive parameters: noise, nonlinearity, and voltage gain. In our analysis these parameters are defined in a way that they include the loading effect of the preceding and following blocks; in other words, the parameters are determined when the block is

P. Sakian et al., *RF-Frontend Design for Process-Variation-Tolerant Receivers*,
Analog Circuits and Signal Processing, DOI 10.1007/978-1-4614-2122-1_2,
© Springer Science+Business Media New York 2012

inside the system. The sensitivity of the overall performance of the receiver to variations of noise, nonlinearity, and gain of building blocks is calculated. In fact, the variations of noise, nonlinearity, and gain of building blocks represent variations in any circuit-level or process-technology-level parameter which can affect the goal function of the receiver (BER as defined in the next section). Therefore, the presented methods include all relevant sources of variability. Since the number of parameters is limited to three for each block, faster computations are possible. On the other hand any part of the receiver which can be characterized by these three parameters can be identified as a building block in the analysis. Therefore, the analysis is flexible in defining the building blocks of the system. To guarantee passing the yield test for all interferers, this analysis is carried out for the specified worst-case interferer.

The presented methods can include RF circuits, analog baseband circuitry, and ADC in the analysis. They are generic and can be used for any circuit topology or process technology, but there are some limitations, as described below:

1. The presented methods are applicable to narrowband systems or to wideband systems, the noise, nonlinearity and gain of which can be represented by an equivalent value across the whole band of operation. Therefore, some types of filters or frequency-dependent components cannot be covered by this analysis.
2. The interferers are assumed to be either completely filtered or to be amplified by almost the same gain as the desired signal is amplified with.
3. The operation bandwidth of the components preceding the mixer is assumed to be narrow enough, so that the harmonics of the LO do not down-convert the noise of these components. In other words, only the fundamental harmonic of the LO down-converts the noise of the preceding stages of the mixer. This assumption allows using the Friis formula to calculate the noise figure of a cascade of stages including a zero-IF mixer. For non-zero-IF mixers additional calculations are done to include the effect of image noise.

Performing the analysis on systems violating any of the above may invalidate the obtained results.

In Sect. 2.1, bit error rate (BER) is defined as the goal function of the receiver and is described as a function of total noise, total nonlinearity, and input impedance of the receiver (for the given type of modulation). In Sect. 2.2, the performance requirements of a typical 60 GHz receiver for indoor applications are described. Total noise and total nonlinearity are described as a function of block-level noise, nonlinearity, and gain in Sect. 2.3 providing a connection between BER and block-level performance parameters. In Sect. 2.4, the relationships derived in Sects. 2.1 and 2.3 are used to determine the sensitivity of the overall performance of the receiver to the performance of the building blocks. It is shown that for all blocks the first order sensitivity to the gain of the block can be nullified, whereas the sensitivity to noise and nonlinearity of all the blocks cannot be minimized simultaneously. In Sect. 2.5, different approaches to minimizing the sensitivities derived in Sect. 2.4 are investigated. In Sect. 2.6 the design of two 60 GHz receivers including and excluding the ADC is explored using different system-level design guidelines presented in this chapter. Section 2.7 brings the conclusion.

2.1   Bit Error Rate, Noise, Gain, and Nonlinearity

Defined as the ratio of erroneous received bits to the total number of received bits, bit error rate (BER) is an essential performance measure for every receiver involved with digital data. The BER in a receiver is, for a given type of modulation, directly determined by the signal to noise-plus-distortion ratio (SNDR). Normally by system-level simulation of the baseband demodulator, one can determine the required SNDR for the desired BER. Having the required SNDR and the specified minimum detectable signal (MDS), the maximum total noise-plus-distortion (NPD) can be calculated by

$$10 \log \left(\frac{NPD}{10^{-3}} \right) = MDS - SNDR \qquad (2.1)$$

where NPD is in watts, MDS is in dBm, and SNDR is measured in dB at the output of the receiver. NPD is selected as the goal function of our analysis and since it is defined as the sum of the noise and nonlinearity distortion it can be described as a function of noise and nonlinearity by

$$NPD = N_{Antenna} + N_{i,tot} + \sum_{q=2}^{\infty} P_{IMDqi,tot} \qquad (2.2)$$

where $N_{Antenna}$ is the noise coming from the antenna, $N_{i,tot}$ is the equivalent input-referred available noise power of the receiver and $P_{IMDqi,tot}$ is the equivalent input-referred available power of in-band qth order intermodulation distortion due to an out-of-band interferer. We have assumed that there is no correlation between noise and distortion and between distortion components of different orders. We have also assumed that the noise and distortion have equal influence on the BER. However, in general that is not the case and the influence of the distortion on the BER depends on many factors such as the modulation type of the interferer. Therefore, if the modulation type is known, (2.2) should be modified in a way that reflects the weight of the distortion. As illustrated in Fig. 2.1, a noisy and nonlinear circuit can be described by an ideal noiseless and linear circuit with equivalent noise and nonlinearity distortions referred to the input. In this case as shown in Fig. 2.1b, input-referred noise and distortions are represented by voltage sources. This representation can also be applied to a zero-IF mixer by considering its RF port as the input and its IF port as the output. In order to describe NPD in terms of equivalent voltages, we need to convert the available power of noise and distortion in (2.2) to their equivalent voltages which results in:

$$NPD =; N_{Antenna} \quad + \quad N_{i,tot} \quad + \quad \sum_{q=2}^{\infty} P_{IMDqi,tot}$$

$$NPD = kTB \quad + \quad \frac{(R_s + R_{in})^2}{4R_s R_{in}^2} \times B\bar{V}_{ni,tot}^2 + \quad \frac{(R_s + R_{in})^2}{4R_s R_{in}^2} \times \sum_{q=2}^{\infty} V_{IMDqi,tot}^2 \qquad (2.3)$$

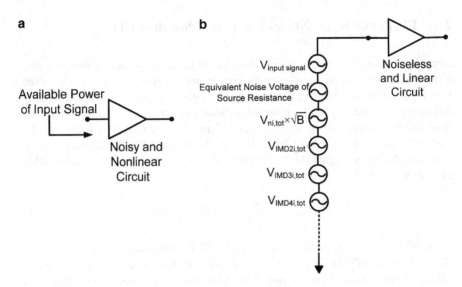

Fig. 2.1 (**a**) A noisy and nonlinear circuit which can serve here both as the total receiver or one of its building blocks, (**b**) corresponding noiseless and linear circuit with equivalent noise and distortion voltage at input

where $k = 1.38 \times 10^{-23}$ J/K is the Boltzmann constant, T is the absolute temperature, B is the effective noise bandwidth, R_{in} is the input impedance, R_s is the source impedance, $V_{ni,tot}$ is the equivalent input-referred noise voltage in V/$\sqrt{Hz}$, and $V_{IMDqi,tot}$ is the equivalent input-referred voltage of qth order intermodulation distortion in volts, as shown in Fig. 2.1b. Without loss of generality and for simplicity, the input impedance of the receiver is assumed to have only a real part.

Once the required $V_{ni,tot}$ and $V_{IMDqi,tot}$ of the whole receiver are determined, overall system specifications including noise factor (NF_{tot}) and the voltage of qth order input intercept point ($V_{IPqi,tot}$) can be calculated:

$$NF_{tot} = \frac{4kT \times R_s + \bar{V}_{ni,tot}^2 \left(\frac{R_{in}+R_s}{R_{in}}\right)^2}{4kT \times R_s} \tag{2.4}$$

$$\frac{1}{V_{IPqi,tot}^{q-1}} = \frac{V_{IMDqi,tot}}{V_{interferer}^q} \tag{2.5}$$

where $V_{interferer}$ is the worst-case out-of-band interferer signal voltage at the input. Substituting $V_{IMDqi,tot}$ from (2.5) in (2.3) yields the NPD as a function of $V_{IPqi,tot}$ and $V_{ni,tot}$:

$$NPD = kTB + \frac{(R_s + R_{in})^2}{4R_s R_{in}^2} \times \left(B\bar{V}_{ni,tot}^2 + \sum_{q=2}^{\infty} \frac{V_{interferer}^{2q}}{V_{IPqi,tot}^{2q-2}} \right) \tag{2.6}$$

The NPD can also be affected by IQ mismatch (only in receivers with I and Q paths) and phase noise. IQ phase or amplitude imbalance results in some cross-talk between I and Q channels. This cross-talk appears as a distortion and affects the NPD. Block-level budgeting of the noise, gain, and nonlinearity does not have a significant (if any) impact on the IQ mismatch. Therefore, if the amount of the IQ cross-talk is known, its impact can be treated as a constant distortion added to the NPD and in this way it can be incorporated in the analysis. Extensive research has been done on IQ mismatch cancellation methods over the past years. Digital IQ imbalance compensation methods are widely used to suppress the impact of IQ mismatch. Details of these methods are beyond the scope of this book.

The phase noise originating from the frequency synthesizer can cause inter-channel and in-band interference. The inter-channel interference can be modeled as a distortion added to the NPD, assuming a worst case adjacent-channel interference. However the impact of in-band interference is signal-dependent and cannot be modeled with just a constant additive distortion, because the effect of phase noise is multiplicative and not additive as thermal noise. Nevertheless, the impact of phase noise is hardly dependent on the noise, gain, and nonlinearity of the building blocks of the receiver. Therefore, the way the block-level budgeting is performed in the receiver can hardly influence the impact of the phase noise on the receiver performance. Minimization of the phase noise impact can be best accomplished in the frequency synthesizer section (e.g. phase-locked loop). In the rest of the chapter the focus will be on the impact of the noise, gain, and nonlinearity distortion of individual blocks of the receiver on the NPD. Therefore, IQ mismatch and phase noise which are rather independent of the block-level budgeting of the receiver will not be addressed any further in this chapter.

2.2 Performance Requirements

The frequency band 57–66 GHz, as allocated by the regulatory agencies in Europe, Japan, Canada, and the United States, can be used for high rate wireless personal area network (WPAN) applications [15]. According to the IEEE 802.15.3C standard, three different modes are possible for the physical layer of such a network: single-carrier mode, high-speed interface mode, and audio/visual mode. The single-carrier mode supports various types of modulation schemes including π/2 QPSK, π/2 8-PSK, π/2 16-QAM, pre-coded MSK, pre-coded GMSK, on-off keying (OOK), and dual alternate mark inversion (DAMI). The high-speed interface mode is designed for non-line-of-sight operation and uses OFDM. The audio/visual mode is also designed for non-line-of-sight operation, uses OFDM, and is considered for uncompressed, high-definition video and audio transport.

The whole band is divided into four channels with center frequencies located at 58.32, 60.480, 62.640, and 64.8 GHz, each with a bandwidth of 2.16 GHz [15]. A transceiver complying with the single-carrier mode should support at least one of the above channels. A transceiver complying with the high-speed interface mode should support at least the channel centered at 60.480 GHz or the one centered at 62.640 GHz. The audio/visual mode is in turn divided into two modes of low data rate (LRP: in the order of 5 Mbps) and high data rate (HRP: in the order of 1 Gbps). A transceiver complying with HRP mode should support at least the channel centered at 60.480 GHz. On the other hand, in the LRP mode the bandwidth of the channels is about 98 MHz, i.e., in each of the above-mentioned channels, three LRP channels, with 98 MHz bandwidth, are defined around the center.

Different physical layer definitions are a result of different possible application demands. For example a kiosk application would require 1.5 Gbps data rate at a 1 m range. This data rate at such a short range can be easily provided by the single-carrier mode with less complexity and thus lower cost compared to physical layer definitions which use OFDM. On the other hand the audio/visual mode is best fitted to uncompressed video streaming applications. An ad-hoc system for connecting computers and devices around a conference table can be best implemented by the high-speed interface mode, because in this case all the devices in the WPAN are expected to have bidirectional, non-line-of-sight high speed, low-latency communication [15].

According to IEEE 802.15.3C standard, the limit for effective isotropic radiated power (EIRP) at this frequency band is 27 dBi for indoor and 40 dBi for outdoor in the United States. EIRP limit in Japan and Australia is 57 and 51.8 dBi, respectively. It is worth reminding that EIRP is the sum of the transmitter output power and its antenna gain.

In the single-carrier mode, the frame error rate must be less than 8%, with a frame payload length of 2,048 octets. The minimum detectable signal varies between -70 and -46 dBm depending on the data rate. The maximum tolerable power level of the incoming signal, which meets the required error rate, is -10 dBm.

In the high-speed interface mode, for a BER of 10^{-6}, the minimum detectable signal varies between -50 and -70 dBm, depending on the required data rate. The maximum tolerable power level of the incoming signal, which meets the required error rate, is -25 dBm.

In the audio/visual mode, a BER of less than 10^{-7} must be met with a bit-stream generated by a special pseudo-random sequence defined in the standard. The audio/visual mode has two different data rate options: high data rate in the order of Gbps and low data rate in the order of Mbps. For low-data-rate and high-data-rate applications the minimum detectable signal of the receiver should be -70 and -50 dBm, respectively. The maximum tolerable power level of the incoming signal is -30 and -24 dBm for low-data-rate and high-data-rate modes, respectively.

2.3 Block-Level Impact

To address the impact of block-level performance on the overall performance, we first start with a rather general case including the image noise and a precise calculation of the total nonlinearity and then we derive a more useful special case.

2.3.1 General Case

In this Section the building blocks of the receiver are described with three parameters including noise, nonlinearity, and gain, as shown in Fig. 2.2a.

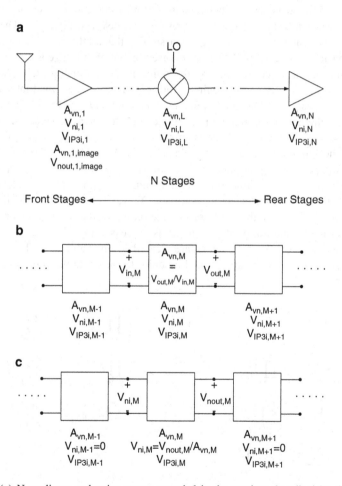

Fig. 2.2 (a) N nonlinear and noisy stages cascaded in the receiver described by their noise, nonlinearity and gain, (b) gain is defined as the ratio between the output voltage of the stage to its input voltage when the stage is in the system, (c) noise of a stage is defined by assuming that all the other stages and antenna are noiseless and observing the noise voltage at the output of the stage

Table 2.1 Some of the notations used in the formulae

Parameter	Notation for: kth stage	Whole receiver
Noise voltage at the output	$V_{nout,k}$	$V_{nout,tot}$
Image noise voltage at the output	$V_{nout,k,image}$	$V_{nout,tot,image}$
Gain: the ratio between output voltage and input voltage at the signal band	$A_{vn,k}$	$A_{vn,tot}$
Image gain: the ratio between output voltage and input voltage at the image band	$A_{vn,k,image}$	$A_{vn,tot,image}$
Equivalent input-referred noise voltage	$V_{ni,k} = \frac{V_{nout,k}}{A_{vn,k}}$	$V_{ni,tot} = \frac{V_{nout,tot}}{A_{vn,tot}}$

The system consists of N stages with at least one mixer as the Lth stage. The gain of each stage is defined as the ratio between the output voltage and the input voltage of each stage when the stage is inside the system, as shown in Fig. 2.2b. The output noise voltage of each stage is observed by assuming that all the other stages and also the antenna are noiseless. Then the input noise voltage of the stage is calculated via dividing the output noise voltage by the gain, as shown in Fig. 2.2c. A similar procedure is used to define the nonlinearities of stages, as described later in more detail. The stages prior to the mixer are characterized by two additional parameters: $A_{vn,k,image}$ is the gain by which the noise voltage of the previous stage (the antenna in case of $k = 1$) at the image band is amplified and $V_{nout,k,image}$ is the noise at the output of the kth stage, generated by the stage itself, residing at the image band. Defining $V_{nout,k}$ as the output noise of the kth stage, generated by the stage itself, $V_{ni,k}$ is defined as $V_{nout,k}$ divided by voltage gain ($A_{vn,k}$). In order to calculate the overall equivalent input-referred noise voltage ($V_{ni,tot}$) as a function of the noise and gain of individual stages, the total noise at the output of the receiver, generated by the receiver itself (and not by the antenna), is expressed in (2.7).

The meanings of all parameters are described in Table 2.1. Two dummy variables $A_{vn,0}$ and $A_{vn,N+1}$ are defined as unity to facilitate the representation of calculations.

$$\bar{V}^2_{nout,tot} = \bar{V}^2_{nout,1} \times \prod_{j=2}^{N} A^2_{vn,j} \qquad + \bar{V}^2_{nout,1,image} \times \prod_{j=2}^{L} A^2_{vn,j,image} \times \prod_{j=L+1}^{N} A^2_{vn,j}$$

$$+ \bar{V}^2_{nout,2} \times \prod_{j=3}^{N} A^2_{vn,j} \qquad + \bar{V}^2_{nout,2,image} \times \prod_{j=3}^{L} A^2_{vn,j,image} \times \prod_{j=L+1}^{N} A^2_{vn,j}$$

$$+ \bar{V}^2_{nout,L-1} \times \prod_{j=L}^{N} A^2_{vn,j} \qquad + \bar{V}^2_{nout,L-1,image} \times A^2_{vn,L,image} \times \prod_{j=L+1}^{N} A^2_{vn,j}$$

$$+ \sum_{k=L}^{N} \bar{V}^2_{nout,k} \times \prod_{j=k+1}^{N+1} A^2_{vn,j} \tag{2.7}$$

Defining two auxiliary variables as in (2.8) and (2.9), replacing $V_{nout,k}$ with $V_{ni,k}A_{vn,k}$, and making some simplifications, the total input-referred noise of the receiver is obtained in (2.10).

$$c_{\mathrm{Im}N,k} = \frac{\bar{V}^2_{nout,k,image}}{\bar{V}^2_{nout,k}} \tag{2.8}$$

$$c_{\mathrm{Im}A,k} = \frac{A^2_{vn,k,image}}{A^2_{vn,k}} \tag{2.9}$$

$$\bar{V}^2_{ni,tot} = \frac{1}{1 + \prod\limits_{j=1}^{L} c_{\mathrm{Im}A,j}} \left[\sum_{k=1}^{L-1} \frac{\left(1 + c_{\mathrm{Im}N,k} \prod\limits_{j=k+1}^{L} c_{\mathrm{Im}A,j}\right) \bar{V}^2_{ni,k}}{\prod\limits_{j=0}^{k-1} A^2_{vn,j}} + \sum_{k=L}^{N} \frac{\bar{V}^2_{ni,k}}{\prod\limits_{j=0}^{k-1} A^2_{vn,j}} \right] \tag{2.10}$$

For obtaining (2.10), the relationship between the effective noise bandwidth (B) in (2.6) and the signal bandwidth (B_{sig}) is defined as follows:

$$B = B_{sig} \times \left(1 + \prod_{j=1}^{L} c_{\mathrm{Im}A,j}\right) \tag{2.11}$$

The contribution of the kth stage to the total noise is calculated from:

$$C_{Noise}(k) = \frac{\bar{V}^2_{ni,k}}{\left(1 + \prod\limits_{j=1}^{L} c_{\mathrm{Im}A,j}\right) \prod\limits_{j=0}^{k-1} A^2_{vn,j}} \times \begin{cases} \left(1 + c_{\mathrm{Im}N,k} \prod\limits_{j=k+1}^{L} c_{\mathrm{Im}A,j}\right), & k < L \\ 1, & k \geq L \end{cases} \tag{2.12}$$

which means that the noise contribution of each stage is inversely proportional to the square of the voltage gain of its preceding stages.

In order to calculate the overall voltage of the qth order input intercept point of the receiver ($V_{IPqi,tot}$) in terms of that of the individual stages, the phase relationships between the nonlinearities and the gains of the stages must be introduced into the calculations.

To start the analysis, the nonlinearity and gain parameters of the receiver and its stages are expressed in phasor form as shown in Table 2.2. The total input-referred voltage of the qth order intermodulation distortion can be expressed in terms of the distortion of the individual stages by (2.13).

Table 2.2 Some of the notations in phasor form

Parameter	Phasor notation for:	
	kth stage	Whole receiver
Input-referred voltage of the qth order intermodulation distortion	$V_{IMDqi,k} \angle \varphi_{IMDqi,k}$	$V_{IMDqi,tot} \angle \varphi_{IMDqi,tot}$
Voltage of the qth order input intercept point	$V_{IPqi,k} \angle \varphi_{IPqi,k}$	$V_{IPqi,tot} \angle \varphi_{IPqi,tot}$
Voltage gain including the loading effect	$A_{vn,k} \angle \varphi_{vn,k}$	$A_{vn,tot} \angle \varphi_{vn,tot}$

$$V_{IMDqi,tot} \angle \varphi_{IMDqi,tot} = \sum_{k=1}^{N} \frac{V_{IMDqi,k} \angle \varphi_{IMDqi,k}}{\prod_{j=0}^{k-1} A_{vn,j} \angle \varphi_{vn,j}} \tag{2.13}$$

Then the distortions can be expressed in terms of intercept points:

$$V_{IMDqi,k} \angle \varphi_{IMDqi,k} = \frac{V_{interferer}^{q} \times \prod_{j=0}^{k-1} A_{vn,j}^{q} \angle q\varphi_{vn,j}}{V_{IPqi,k}^{q-1} \angle (q-1)\varphi_{IPqi,k}} \tag{2.14}$$

Therefore the total intercept point can be expressed in terms of the intercept points of the individual stages, by substituting the distortions from (2.14) into (2.13):

$$\frac{1}{V_{IPqi,tot}^{q-1} \angle (q-1)\varphi_{IPqi,tot}} = \sum_{k=1}^{N} \left(\frac{\prod_{j=0}^{k-1} A_{vn,j}^{q-1}}{V_{IPqi,k}^{q-1}} \angle (q-1) \left(\sum_{j=0}^{k-1} \varphi_{vn,j} - \varphi_{IPqi,k} \right) \right) \tag{2.15}$$

or

$$\frac{1}{\left| V_{IPqi,tot} \right|^{q-1}} = \sqrt{ \left(\sum_{k=1}^{N} \left(\frac{\prod_{j=0}^{k-1} A_{vn,j}^{q-1}}{V_{IPqi,k}^{q-1}} \cos\left((q-1) \left(\sum_{j=0}^{k-1} \varphi_{vn,j} - \varphi_{IPqi,k} \right) \right) \right) \right)^{2} + \left(\sum_{k=1}^{N} \left(\frac{\prod_{j=0}^{k-1} A_{vn,j}^{q-1}}{V_{IPqi,k}^{q-1}} \sin\left((q-1) \left(\sum_{j=0}^{k-1} \varphi_{vn,j} - \varphi_{IPqi,k} \right) \right) \right) \right)^{2} } \tag{2.16}$$

2.3.2   Special Case: Zero-IF Receiver and Worst-Case Nonlinearity

In the special case of a zero-IF receiver with worst-case nonlinearity superposition, the above relationships can be simplified substantially. This special case is the focus of the most of the analysis in the rest of this work.

In a zero-IF mixer with a complex mixer and I/Q signal paths, (2.10), (2.11), and (2.12) simplify to (2.17), (2.18), and (2.19).

$$\bar{V}_{ni,tot}^2 = \sum_{k=1}^{N} \frac{\bar{V}_{ni,k}^2}{\prod\limits_{j=0}^{k-1} A_{vn,j}^2} \tag{2.17}$$

$$B = B_{sig} \tag{2.18}$$

$$C_{Noise}(k) = \frac{\bar{V}_{ni,k}^2}{\prod\limits_{j=0}^{k-1} A_{vn,j}^2} \tag{2.19}$$

If the intermodulation distortions of the consecutive stages are added in-phase resulting in a worst-case scenario, the expression for the intercept point in (2.15) simplifies to below [12]:

$$\frac{1}{V_{IPqi,tot}^{q-1}} = \sum_{k=1}^{N} \frac{\prod\limits_{j=0}^{k-1} A_{vn,j}^{q-1}}{V_{IPqi,k}^{q-1}} \tag{2.20}$$

The above relationships are based on the assumption that the interferer and the in-band signal are amplified with almost the same gain. However, if the interferer is far from the in-band signal in the frequency domain it can be filtered at the input of the receiver such that it generates no distortion.

Based on (2.20), the contribution of the kth stage to the total qth order nonlinearity distortion is equal to:

$$C_{Distortion,q}(k) = \frac{\prod\limits_{j=0}^{k-1} A_{vn,j}^{q-1}}{V_{IPqi,k}^{q-1}} \tag{2.21}$$

which means that the nonlinearity distortion contribution of each stage is directly proportional to the combined voltage gain of its preceding stages. This in fact creates a trade-off in defining the gain of stages. The parameters C_{Noise} and $C_{Distortion}$ will play a central role in the rest of our analysis.

2.4   Sensitivity to Block-Level Performance

In analogy with the previous section, the sensitivities of the overall performance to the block-level performance are calculated in both general and specific case.

2.4.1   General Case

To find a block-level budgeting for optimum robustness of the receiver to variability of block-level performance, one has to minimize the sensitivity of the total performance to the performance of individual blocks. In fact, one way to make the receiver robust to process variations is to make it robust to performance degradations of its building blocks. In this section, the sensitivity of total NPD to the noise, nonlinearity, and gain of individual blocks is calculated. The normalized single-point sensitivity of F(x) to the variable x is defined by the following operator [8]:

$$S_x^{F(x)} \triangleq \frac{\partial F(x)}{\partial x} \times \frac{x}{F(x)} \tag{2.22}$$

which should be calculated at the selected nominal values of x and F(x). As mentioned earlier, to achieve a certain BER a specific NPD requirement has to be met. Therefore, variations of NPD can cause variations in BER. The variations of NPD can be described as a function of the variation of the performance parameters of the individual stages:

$$\Delta NPD = \sum_{k=1}^{N} \frac{\partial NPD}{\partial A_{vn,k}} \Delta A_{vn,k} + \sum_{k=1}^{N} \frac{\partial NPD}{\partial V_{ni,k}} \Delta V_{ni,k} + \sum_{q=2}^{\infty} \sum_{k=1}^{N} \frac{\partial NPD}{\partial V_{IPqi,k}} \Delta V_{IPqi,k} + \frac{\partial NPD}{\partial Z_{in}} \Delta Z_{in}$$

$$\tag{2.23}$$

The random variables $\Delta A_{vn,k}$, $\Delta V_{ni,k}$, $\Delta V_{IPqi,k}$, and ΔZ_{in} which represent the performance variations of the individual stages and the input impedance are usually correlated. However, regardless of the amount of correlation between these random variables, one can minimize ΔNPD by minimizing (ideally nullifying) the derivatives in (2.23), which are proportional to the sensitivity functions. Furthermore, the random variables $\Delta A_{vn,k}$, $\Delta V_{ni,k}$, and $\Delta V_{IPqi,k}$ are in general different for different stages (i.e., different values of k). If they are significantly larger for a specific stage, it is advisable to focus on reducing the sensitivity of the total performance to the performance of that stage. However, such knowledge requires circuit-level information about each stage which may be achieved during the circuit design. Therefore, in this analysis, we attempt to reduce all the sensitivities, assuming that the random variables $\Delta A_{vn,k}$, $\Delta V_{ni,k}$, and $\Delta V_{IPqi,k}$ are in the same order for different stages. The sensitivity functions of NPD to performance parameters of each stage ($V_{ni,k}$, $V_{IPqi,k}$, and $A_{vn,k}$) are listed in (2.25), (2.26), and (2.27) respectively. The derivatives are taken using the chain rule applied to (2.6), (2.10), and (2.16). To simplify the equations, an auxiliary variable is defined in (2.24).

$$R_{eq} \triangleq \frac{4R_s R_{in}^2}{(R_s + R_{in})^2} \tag{2.24}$$

$$S_{V_{ni,k}^2}^{NPD} = \frac{B}{\left(1 + \prod\limits_{j=1}^{L} c_{ImA,j}\right) R_{eq} \times NPD} \times \frac{\bar{V}_{ni,k}^2}{\prod\limits_{j=0}^{k-1} A_{vn,j}^2}$$

$$\times \begin{cases} \left(1 + c_{ImN,k} \prod\limits_{j=k+1}^{L} c_{ImA,j}\right), & k<L \\[2ex] 1, & k \geq L \end{cases} \tag{2.25}$$

$$S_{\left(\frac{1}{V_{IPqi,k}^{q-1}}\right)}^{NPD} = \frac{(q-1)V_{interferer}^{2q}}{R_{eq} \times NPD} \times \frac{\prod\limits_{j=0}^{k-1} A_{vn,j}^{q-1}}{V_{IPqi,k}^{q-1}}$$

$$\times \left[\cos\left((q-1)\left(\sum_{j=0}^{k-1} \varphi_{vn,j} - \varphi_{IPqi,k}\right)\right) \sum_{m=1}^{N} \left(\frac{\prod\limits_{j=0}^{m-1} A_{vn,j}^{q-1}}{V_{IPqi,m}^{q-1}} \cos\left((q-1)\left(\sum_{j=0}^{m-1} \varphi_{vn,j} - \varphi_{IPqi,m}\right)\right)\right) \right.$$

$$\left. + \sin\left((q-1)\left(\sum_{j=0}^{k-1} \varphi_{vn,j} - \varphi_{IPqi,k}\right)\right) \sum_{m=1}^{N} \left(\frac{\prod\limits_{j=0}^{m-1} A_{vn,j}^{q-1}}{V_{IPqi,m}^{q-1}} \sin\left((q-1)\left(\sum_{j=0}^{m-1} \varphi_{vn,j} - \varphi_{IPqi,m}\right)\right)\right) \right]$$

$$\tag{2.26}$$

It is clear from (2.25), (2.26), and (2.27) that the sensitivities to the noise and nonlinearity of the stages cannot be made zero, whereas the sensitivity to the gain of individual stages can be made zero.

$$
S_{A_{vn,k}}^{NPD} = \begin{cases}
\dfrac{2}{R_{eq} \times NPD} \times \displaystyle\sum_{q=2}^{\infty} \left[\left(\sum_{m=1}^{N} \left(\dfrac{(q-1)V_{interferer}^{2q} \prod_{j=0}^{m-1} A_{vn,j}^{q-1}}{V_{IPqi,m}^{q-1}} \cos\left((q-1)\left(\sum_{j=0}^{m-1} \varphi_{vn,j} - \varphi_{IPqi,m} \right) \right) \right) \right) \right. \\[2em]
\left(\displaystyle\sum_{m=k+1}^{N} \left(\dfrac{\prod_{j=0}^{m-1} A_{vn,j}^{q-1}}{V_{IPqi,m}^{q-1}} \cos\left((q-1)\left(\sum_{j=0}^{m-1} \varphi_{vn,j} - \varphi_{IPqi,m} \right) \right) \right) \right) \\[2em]
+ \left(\displaystyle\sum_{m=1}^{N} \left(\dfrac{(q-1)V_{interferer}^{2q} \prod_{j=0}^{m-1} A_{vn,j}^{q-1}}{V_{IPqi,m}^{q-1}} \sin\left((q-1)\left(\sum_{j=0}^{m-1} \varphi_{vn,j} - \varphi_{IPqi,m} \right) \right) \right) \right) \\[2em]
\left. \left(\displaystyle\sum_{m=k+1}^{N} \left(\dfrac{\prod_{j=0}^{m-1} A_{vn,j}^{q-1}}{V_{IPqi,m}^{q-1}} \sin\left((q-1)\left(\sum_{j=0}^{m-1} \varphi_{vn,j} - \varphi_{IPqi,m} \right) \right) \right) \right) \right], \quad k < L-1 \\[3em]
-\dfrac{2}{R_{eq} \times NPD} \displaystyle\sum_{j=k+1}^{L-1} \dfrac{B\left(1 + c_{ImN,j} \prod_{m=j+1}^{L} c_{ImA,m}\right) \bar{V}_{ni,j}^2}{\left(1 + \prod_{j=1}^{L} c_{ImA,j}\right) \prod_{m=0}^{j-1} A_{vn,m}^2} - \dfrac{2}{R_{eq} \times NPD} \sum_{j=L}^{N} \dfrac{B \bar{V}_{ni,j}^2}{\left(1 + \prod_{j=1}^{L} c_{ImA,j}\right) \prod_{m=0}^{j-1} A_{vn,m}^2} \\[3em]
\dfrac{2}{R_{eq} \times NPD} \times \displaystyle\sum_{q=2}^{\infty} \left[\left(\sum_{m=1}^{N} \left(\dfrac{(q-1)V_{interferer}^{2q} \prod_{j=0}^{m-1} A_{vn,j}^{q-1}}{V_{IPqi,m}^{q-1}} \cos\left((q-1)\left(\sum_{j=0}^{m-1} \varphi_{vn,j} - \varphi_{IPqi,m} \right) \right) \right) \right) \right. \\[2em]
\left(\displaystyle\sum_{m=k+1}^{N} \left(\dfrac{\prod_{j=0}^{m-1} A_{vn,j}^{q-1}}{V_{IPqi,m}^{q-1}} \cos\left((q-1)\left(\sum_{j=0}^{m-1} \varphi_{vn,j} - \varphi_{IPqi,m} \right) \right) \right) \right) \\[2em]
+ \left(\displaystyle\sum_{m=1}^{N} \left(\dfrac{(q-1)V_{interferer}^{2q} \prod_{j=0}^{m-1} A_{vn,j}^{q-1}}{V_{IPqi,m}^{q-1}} \sin\left((q-1)\left(\sum_{j=0}^{m-1} \varphi_{vn,j} - \varphi_{IPqi,m} \right) \right) \right) \right) \\[2em]
\left. \left(\displaystyle\sum_{m=k+1}^{N} \left(\dfrac{\prod_{j=0}^{m-1} A_{vn,j}^{q-1}}{V_{IPqi,m}^{q-1}} \sin\left((q-1)\left(\sum_{j=0}^{m-1} \varphi_{vn,j} - \varphi_{IPqi,m} \right) \right) \right) \right) \right] \\[2em]
-\dfrac{2}{R_{eq} \times NPD} \displaystyle\sum_{j=k+1}^{N} \dfrac{B \bar{V}_{ni,j}^2}{\left(1 + \prod_{j=1}^{L} c_{ImA,j}\right) \prod_{m=0}^{j-1} A_{vn,m}^2}, \\[2em]
k \geq L-1
\end{cases}
$$

$$(2.27)$$

2.4.2 Special Case: Zero-IF Receiver and Worst-Case Nonlinearity

If the receiver is zero-IF with a complex mixer and I/Q signal paths in IF and the design is done for worst case scenario in which the intermodulation distortions of the individual stages are assumed to add up in phase, (2.25), (2.26), and (2.27) are simplified to (2.28), (2.29), and (2.30).

$$S^{NPD}_{\bar{V}^2_{ni,k}} = \frac{B}{R_{eq} \times NPD} \times \frac{\bar{V}^2_{ni,k}}{\prod\limits_{j=0}^{k-1} A^2_{vn,j}} \tag{2.28}$$

$$S^{NPD}_{\left(\frac{1}{V^{q-1}_{IPqi,k}}\right)} = \frac{2V^{2q}_{interferer}}{R_{eq} \times V^{q-1}_{IPqi,tot} \times NPD} \times \frac{\prod\limits_{j=0}^{k-1} A^{q-1}_{vn,j}}{V^{q-1}_{IPqi,k}} \tag{2.29}$$

$$S^{NPD}_{A_{vn,k}} = \frac{2}{R_{eq} \times NPD}$$

$$\times \left(\sum_{q=2}^{\infty} \left(\frac{(q-1)V^{2q}_{interferer}}{V^{q-1}_{IPqi,tot}} \times \sum_{j=k+1}^{N} \frac{\prod\limits_{m=1}^{j-1} A^{q-1}_{vn,m}}{V^{q-1}_{IPqi,j}} \right) - \sum_{j=k+1}^{N} \frac{B\bar{V}^2_{ni,j}}{\prod\limits_{m=1}^{j-1} A^2_{vn,m}} \right) \tag{2.30}$$

The second order sensitivity of NPD to block-level gain is proportional to the second order derivative of NPD with respect to block-level gains as described in (2.31) and (2.32):

$$\begin{cases} \frac{\partial^2 NPD}{\partial A_{vn,k}\partial A_{vn,p}} = \frac{2R^{-1}_{eq}}{A_{vn,k}A_{vn,p}} \times \left(\sum_{q=2}^{\infty} \left((q-1)^2 V^{2q}_{interferer} \times \sum_{j=p+1}^{N} \frac{\prod\limits_{m=1}^{j-1} A^{q-1}_{vn,m}}{V^{q-1}_{IPqi,j}} \right. \right. \\ \left. \left. \times \left(\frac{1}{V^{q-1}_{IPqi,tot}} + \sum_{j=k+1}^{N} \frac{\prod\limits_{m=1}^{j-1} A^{q-1}_{vn,m}}{V^{q-1}_{IPqi,j}} \right) \right) + 2 \times \sum_{j=p+1}^{N} \frac{B\bar{V}^2_{ni,j}}{\prod\limits_{m=1}^{j-1} A^2_{vn,m}} \right) \\ (p>k) \end{cases} \tag{2.31}$$

$$\frac{\partial^2 NPD}{\partial A_{vn,k}^2} = \frac{2R_{eq}^{-1}}{A_{vn,k}^2} \times \left(\sum_{q=2}^{\infty} \left(V_{interferer}^{2q} \left((q-1)^2 \left(\sum_{j=k+1}^{N} \frac{\prod_{m=1}^{j-1} A_{vn,m}^{q-1}}{V_{IPqi,j}^{q-1}} \right)^2 \right. \right. \right.$$

$$\left. \left. \left. + \frac{(q-1)(q-2)}{V_{IPqi,tot}^{q-1}} \times \sum_{j=k+1}^{N} \frac{\prod_{m=1}^{j-1} A_{vn,m}^{q-1}}{V_{IPqi,j}^{q-1}} \right) \right) + 3 \sum_{j=k+1}^{N} \frac{B\bar{V}_{ni,j}^2}{\prod_{m=1}^{j-1} A_{vn,m}^2} \right) \qquad (2.32)$$

which are valid for ($1 \le k < N$).

An inspection of (2.19) and (2.28), and in general case (2.12) and (2.25), shows that the sensitivity to the noise performance of each stage is proportional to its contribution to the total noise of the receiver. A similar inspection of (2.21) and (2.29) reveals that the sensitivity to the qth order nonlinearity of each component is proportional to its contribution to the total qth order nonlinearity. Therefore, for a given total noise and total nonlinearity, reducing the sensitivity to noise/nonlinearity of one stage results in increased sensitivity to noise/nonlinearity of other stages. On the other hand, the sensitivity to gain of each block can be set to zero as implied by (2.27) or (2.30). Furthermore, according to (2.31) and (2.32), the second order sensitivity to block-level gains cannot be nullified but can be reduced by lowering the contribution of the rear stages to the total noise and nonlinearity.

One solution for zeroing the first order sensitivity of NPD to gains in (2.30) is given as below:

$$\frac{\bar{V}_{ni,k}^2}{\bar{V}_{ni,k-1}^2} = \alpha_{k-1} A_{vn,k-1}^2 \qquad (2.33)$$

$$\frac{V_{IPqi,k}^{q-1}}{V_{IPqi,k-1}^{q-1}} = \frac{1}{\alpha_{k-1}} A_{vn,k-1}^{q-1} \qquad (2.34)$$

$$B\bar{V}_{ni,tot}^2 = \sum_{q=2}^{\infty} \frac{(q-1)V_{interferer}^{2q}}{V_{IPqi,tot}^{2(q-1)}} \qquad (2.35)$$

where α_k ($1 \le k < N$) is a parameter that must be chosen by the designer and we call it contribution factor of a stage, because it determines the ratio of the noise and nonlinearity distortion contribution of each stage to that of its following stage. To nullify (2.30) for every stage, (2.33), (2.34), and (2.35) must be satisfied for ($1 < k \le N$). In case of dominance of third order nonlinearity (2.35) simplifies to:

$$B\bar{V}_{ni,tot}^2 = \frac{2V_{interferer}^6}{V_{IP3i,tot}^4} = 2V_{IMD3i,tot}^2 \qquad (2.36)$$

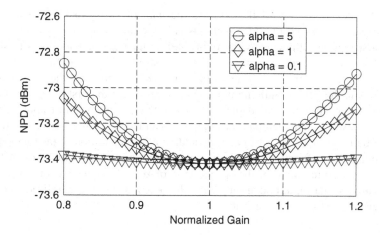

Fig. 2.3 NPD versus normalized variation of the gain of the first stage for different values of α

which means that the third order intermodulation distortion must be 3 dB below the level of the noise coming from the receiver itself.

According to (2.31) and (2.32), the second order sensitivity of NPD to the gain of front stages is bigger than that of rear stages. For a system which satisfies (2.33), (2.34), and (2.35), the reduction of the second order sensitivity to the gain of front stages requires their α to be smaller than unity, resulting in a lower contribution of rear stages to total noise and nonlinearity. In fact if α is smaller than unity for every stage, the contribution of each stage to the total noise and nonlinearity, as quantified by (2.12), (2.19), and (2.21), will be smaller than that of its previous stage. It is worth mentioning that a higher contribution of a stage to the total noise (or nonlinearity distortion) does not necessarily mean that it is noisier (or more nonlinear), because the contribution of one stage to total noise (or total nonlinearity distortion) is not only a function of the noise (or nonlinearity) of the stage itself but also a function of the gain of its previous stages, as described by (2.12), (2.19), and (2.21). Figure 2.3 shows the NPD of a receiver as a function of the gain of its first stage (normalized to its nominal value) for different values of α. It shows the significant impact of α on the second order sensitivity, with smaller values of α yielding smaller second-order sensitivity to the gain of the LNA.

2.5 Design for Robustness

The equations derived in Sect. 2.4 are insightful for both analysis and synbook of a receiver. From the analysis perspective, they can determine the most sensitive building blocks of the receiver. From the synbook perspective they can be used to develop design approaches for minimum sensitivity to variability of building blocks. In this section, three design approaches are investigated. In all of them the first order sensitivity of the NPD to the gain of each block is set to zero. The difference between the three methods is in the different values chosen for α. In the first approach, the sensitivities of the NPD to the noise and nonlinearity of the individual stages are all the same while the sensitivity of the NPD to the gain of each block is set to zero, by setting α to 1. In the second approach, the second order sensitivity to the gain of building blocks is reduced while the first order sensitivity to gain is kept at zero, by keeping α smaller than 1. In the third approach an optimum-power design method is described [12] and the first order sensitivity to the gain of building blocks is set to zero by adding a new condition to the method; in this method α is a function of the power consumption of the components. All the approaches are presented for the special case of zero-IF receiver with worst case nonlinearity scenario, although they have the potential to be generalized to other receiver architectures and scenarios.

2.5.1 Constant-Sensitivity Approach

For a given total noise and total nonlinearity, reducing the sensitivity to noise/nonlinearity of one stage results in increased sensitivity to noise/nonlinearity of other stages, because reducing the contribution of one stage to total noise or nonlinearity, (2.19) and (2.21), increases the contribution of other stages. As a result, by keeping sensitivities equal for all the stages one can avoid extra-sensitive nodes in the receiver. Forcing the sensitivity to noise in (2.28) to be equal for all k ($1 \leq k \leq N$) and using (2.17) and (2.19) yields:

$$C_{Noise}(k) = \frac{\bar{V}_{ni,k}^2}{\prod\limits_{j=0}^{k-1} A_{vn,j}^2} = \frac{\bar{V}_{ni,tot}^2}{N} \tag{2.37}$$

which means that the input-referred noise voltage of each stage must be better than that of its following stage by a factor of its voltage gain:

$$\frac{\bar{V}_{ni,k}^2}{\bar{V}_{ni,k-1}^2} = A_{vn,k-1}^2 \tag{2.38}$$

Forcing the sensitivity to nonlinearity in (2.29) to be constant for all k ($1 \leq k \leq N$) and using (2.20) and (2.21) yields:

$$C_{Distortion,q}(k) = \frac{\prod\limits_{j=0}^{k-1} A_{vn,j}^{q-1}}{V_{IPqi,k}^{q-1}} = \frac{1}{N \times V_{IPqi,tot}^{q-1}} \tag{2.39}$$

which means that the qth order nonlinearity performance of each stage must be better than that of its preceding stage by a factor of voltage gain of its preceding stage:

$$\frac{V_{IPqi,k}^{q-1}}{V_{IPqi,k-1}^{q-1}} = A_{vn,k-1}^{q-1} \tag{2.40}$$

In the constant-sensitivity approach, (2.37) and (2.39) guarantee that the sensitivity to noise and nonlinearity of all the stages are the same and the contribution of all the stages to total noise and nonlinearity are equal.

According to (2.38) and (2.40) two conditions for zeroing the first order sensitivity of NPD to the gain, (2.33) and (2.34), are automatically satisfied in this approach with α equal to unity for all the stages. Consequently, the designer has to satisfy (2.35) to nullify the first order sensitivity to the gain.

Therefore, in this approach first order sensitivity to gain of each stage is set to zero and the sensitivity to noise and nonlinearity of all stages is equal.

2.5.2 Reduced-Second-Order-Sensitivity Approach

In this approach in addition to nullifying the first order sensitivity to block-level gains in (2.30), the second order sensitivity to the gain of front stages in (2.31) and (2.32) is reduced. To achieve that, α must be smaller than unity for every stage, resulting in:

$$\frac{\overline{V}_{ni,k}^2}{\overline{V}_{ni,k-1}^2} < A_{vn,k-1}^2 \tag{2.41}$$

$$\frac{V_{IPqi,k}^{q-1}}{V_{IPqi,k-1}^{q-1}} > A_{vn,k-1}^{q-1} \tag{2.42}$$

In contrast to the constant-sensitivity approach, in the reduced second order sensitivity approach there is no unique solution for zeroing (2.30); i.e., choosing any α smaller than unity is sufficient. In fact making α smaller for each stage can further reduce the value of the second order sensitivity to the gain of that stage. Theoretically, there is no lower limit for α and setting α to zero yields the lowest second-order sensitivity; however, in practice obtaining very low values of α is difficult because it demands highly linear and low-noise components which are either impractical to implement or entail extremely high power consumptions.

In this approach more attention is paid to the sensitivity of the NPD to the gain of individual blocks and no constraints are defined for sensitivity to the noise and nonlinearity of building blocks, because reducing the sensitivity of the NPD to the noise or nonlinearity of one stage increases the sensitivity of the NPD to the noise and nonlinearity of other stages, as described in the last subsection. Therefore, the overall impact of adjusting the sensitivities to block-level noise and nonlinearity on the robustness is not expected to be significant. Consequently, the reduced-second-order-sensitivity approach is expected to provide more robustness to process variations as compared to the constant-sensitivity approach.

As a side effect of reducing the second order sensitivity of the NPD to the gain of the stages, the sensitivity to noise and nonlinearity of front stages (such as LNA) is increased in this approach whereas the sensitivity to noise and nonlinearity of rear stages (like baseband filters) is reduced.

2.5.3 Optimum-Power Design

In optimum-power design as suggested in Ref. [12], a linear relationship is assumed between power consumption and dynamic range of the components:

$$P = \frac{V_{IP3i}^2}{\bar{V}_{ni}^2} P_C \tag{2.43}$$

where P is the power consumption of the component and P_C is the proportionality constant and is called power coefficient. This assumption is valid in many RF components in specific operating regions [12]. Therefore, the resulting method is applicable solely in those regions. In this book, the optimum-power method is used as a means of comparison with the methods developed in this work so that we can verify that the methods aiming at robustness may also satisfy the requirements of optimum power design. Optimum-power methodology proposes a block-level budgeting plan which results in minimum power consumption for a given total noise and nonlinearity requirement [12]:

$$C_{Noise}(k) = \frac{\bar{V}_{ni,k}^2}{\prod\limits_{j=0}^{k-1} A_{vn,j}^2} = \bar{V}_{ni,tot}^2 \times \frac{\sqrt[3]{P_{Ck}}}{\sum\limits_{m=1}^{N} \sqrt[3]{P_{Cm}}} \tag{2.44}$$

$$C_{Distortion,3}(k) = \frac{\prod\limits_{j=0}^{k-1} A_{vn,j}^2}{V_{IP3i,k}^2} = \frac{1}{V_{IP3i,tot}^2} \times \frac{\sqrt[3]{P_{Ck}}}{\sum\limits_{m=1}^{N} \sqrt[3]{P_{Cm}}} \tag{2.45}$$

where P_{Ck} is the power coefficient of the kth stage and the total power consumption after optimization is obtained from (2.46). In this approach, third order nonlinearity is assumed to be the dominant source of distortion and the other orders of nonlinearity are neglected.

$$P_{tot} = \frac{V_{IP3i,tot}^2}{V_{ni,tot}^2} \times \left(\sum_{m=1}^{N} \sqrt[3]{P_{Cm}} \right)^3 \qquad (2.46)$$

In the special case that the power coefficients of all blocks are equal, (2.44) and (2.45) simplify to (2.37) and (2.39), respectively and the optimum-power design gives the same results as constant-sensitivity design.

However, the guideline proposed in Ref. [12] does not determine the total noise and total nonlinearity of the receiver. As was mentioned in Sect. 2.1, the required BER determines the required noise-plus-distortion (NPD) but does not specify the contribution of the noise or nonlinearity to the total NPD. Thus the designer needs to specify $V_{IP3i,tot}$ and $V_{ni,tot}$ in such a way that the total power consumption in (2.46) is minimized and the required NPD in (2.3) is satisfied. It is proven that minimum power is achieved when the condition of (2.36) is met (see Appendix A).

An inspection of (2.44) and (2.45) reveals that in this approach the noise and nonlinearity of consecutive stages have the following relationships:

$$\frac{V_{ni,k}^2}{V_{ni,k-1}^2} = A_{vn,k-1}^2 \times \frac{\sqrt[3]{P_{Ck}}}{\sqrt[3]{P_{Ck-1}}} \qquad (2.47)$$

$$\frac{V_{IPqi,k}^{q-1}}{V_{IPqi,k-1}^{q-1}} = A_{vn,k-1}^{q-1} \times \frac{\sqrt[3]{P_{Ck-1}}}{\sqrt[3]{P_{Ck}}} \qquad (2.48)$$

which mean that this approach also satisfies the first two conditions for zeroing the first order sensitivity of NPD to the gain of stages as described in (2.33) and (2.34) and α is equal to the third root of the ratio of the power coefficients. Therefore satisfying (2.36) is also necessary for zeroing the first order sensitivity to the gains. Thus, selecting the total noise and nonlinearity of the receiver for minimum power leads to the nullification of the first order sensitivity of NPD to block-level gains. This is an important advantage of the optimum power design.

Based on (2.44), (2.45) and (2.28), (2.29), the sensitivity to the noise and nonlinearity of each block is directly proportional to the third root of its power coefficient meaning that the overall system performance is more sensitive to power-hungry blocks.

In the rest of the chapter, the optimum-power approach refers to the herewith modified version of the method introduced in Ref. [12] which also satisfies the condition of (2.36).

In the course of development of this method it is assumed that the noise and nonlinearity performance of an RF circuit can be improved linearly and indefinitely by just increasing the power consumption and that for constant power consumption the noise and linearity performance can be traded for each other [12]. Due to the limited range of validity of these assumptions, the method should be used with caution.

Table 2.3 Summary of the three explained approaches

	Constant sensitivity	Reduced second order sensitivity	Optimum power
Contribution factor (α_k)	=1	<1	$= \frac{\sqrt[3]{P_{Ck+1}}}{\sqrt[3]{P_{Ck}}}$
First order sensitivity to the gain of stages	0	0	0
Second order sensitivity to the gain of front stages	High	Low	High if rear stages are power hungry/ low if rear stages are low-power
Sensitivity to noise and nonlinearity of each stage	Constant	High for front stages/low for rear stages	Proportional to third root of power coefficient

2.5.4 Summary and Discussion

The three approaches studied in this section, are summarized in Table 2.3. Clearly, these approaches should be compared when their specified NPDs are equal. According to the second and last row of the table, the optimum power design only satisfies the necessary conditions for reduced second-order sensitivity design, if the power coefficients of the rear stages are smaller than those of the front stages. Alternatively, the optimum-power design only gives the same specifications of the constant-sensitivity approach, if the power coefficients of all the stages are equal.

In the special case that the power coefficient of each block is smaller than that of its previous stage, the α of the optimum-power design is smaller than unity and the optimum-power design meets the conditions of the reduced-second-order sensitivity approach. Accordingly, the second-order sensitivity of the NPD to the gain of a stage is higher if its following stages have larger power coefficients. This is because in optimum-power design, the blocks with larger power coefficients have higher contribution to the total noise (C_{Noise}) and nonlinearity distortion ($C_{Distortion}$) and that can be verified by inspection of (2.19) and (2.21) and (2.47) and (2.48). In other words, for a stage which is followed by another one with a larger power coefficient, the value of α is larger than unity, contradicting the requirements of the reduced-second-order-sensitivity approach. In fact, optimum-power design does not put any constraint on the position of large-power-coefficient blocks, but having this sort of blocks at the front stages of the receiver would let the optimum-power design produce the same specifications as suggested by the reduced-second-order-sensitivity approach.

As a conclusion, the reduced-second-order-sensitivity approach is the ideal solution in terms of improving the robustness of a receiver to process variations. However, it is not always a practical solution considering its implications for the power consumption. If the large-power-coefficient blocks are located in the front stages of the receiver, the reduced second-order sensitivity approach can be used without excessive increase in the power consumption. However, if the large-power-coefficient blocks are in the rear stages of the receiver, the optimum-power method is preferred as it meets the requirement for zeroing the first-order sensitivity to the gain of all the stages while yielding the minimum power consumption.

2.6 Case Study

In this section we apply the design approaches described in Sect. 2.5 to 60 GHz zero-IF receivers. First, we investigate the case of a receiver without analog-to-digital converter (ADC) and then we study another receiver which includes an ADC.

2.6.1 Receiver Without ADC

The ADC-less 60 GHz zero-IF receiver shown in Fig. 2.4 is considered, to perform a comparison between different design approaches. A non-optimum design based on a tested 60 GHz LNA and mixer is also included in the comparison. The power coefficients, used to estimate the power consumption in each case, and the block-level gains are extracted from simulated and measured circuits [16, 17].

The requirements of the overall system are an SNDR of 12 dB and an MDS of −61.4 dBm which result in an NPD of −73.4 dBm. The tolerable interference power at the input of the receiver is −33 dBm and the RF bandwidth is 2 GHz. For the sake of practicality, these requirements are based on state-of-the-art 60 GHz receivers [18]. Assuming that third order intermodulation distortions are the dominant part of nonlinearities, (2.36), (2.3), (2.4), and (2.5) can be used to find the required noise figure and IIP3 which are 6 dB and −10 dBm, respectively. Block-level specifications obtained from each of the four design approaches are listed in Table 2.4. The input impedance of the receiver is assumed to be 50 Ω for all cases. All four designs satisfy the noise figure and IP3 requirements. The power consumption is estimated by [12]:

$$P_{tot} = \sum_{m=1}^{N} \frac{V_{IP3i,m}^2}{V_{ni,m}^2} P_{Cm} \tag{2.49}$$

The variation of NPD as a function of the variations in the gain of the LNA, for the four approaches, is illustrated in Fig. 2.5. The gain of the LNA is varied by ±20% around its nominal value.

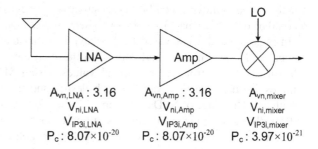

Fig. 2.4 Three-stage ADC-less receiver used in the case study

Table 2.4 Block-level specifications of the four design approaches

Parameter	Constant sensitivity	Reduced second-order sensitivity	Optimum power	Non-optimum
$V_{ni,LNA}^2$ (V^2/Hz)	2.06×10^{-19}	3.599×10^{-19}	2.6×10^{-19}	2.896×10^{-19}
$V_{ni,Amp}^2$ (V^2/Hz)	2.06×10^{-18}	1.542×10^{-18}	2.6×10^{-18}	1.274×10^{-18}
$V_{ni,Mixer}^2$ (V^2/Hz)	2.06×10^{-17}	1.028×10^{-17}	9.56×10^{-18}	1.995×10^{-17}
$V_{IP3i,LNA}^2$ (V^2)	0.015	0.0086	0.0118	0.1256
$V_{IP3i,Amp}^2$ (V^2)	0.15	0.2000	0.118	0.32
$V_{IP3i,Mixer}^2$ (V^2)	1.5	3.0001	3.2	0.621
Estimated power (mW)	12	13.6	8.6	55

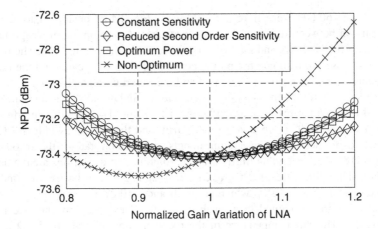

Fig. 2.5 NPD versus normalized variation of LNA gain for the four different designs

Except the non-optimum design, all the approaches result in zero first-order sensitivity of the NPD to the gain of the LNA, whereas the reduced-second-order-sensitivity design shows minimum variation of NPD with respect to gain deviations in the LNA. In this example the second-order sensitivity to gain in optimum-power design is not much higher than the one in reduced-second-order-sensitivity approach, because the power coefficient of the final stage is very low and therefore its noise and nonlinearity contribution is low. Please note that because only the gain of one stage is varied in Fig. 2.5, the variations of NPD are small. The value of α in the case of reduced-second-order sensitivity approach is 0.43. If smaller values of α could be obtained, smaller second-order sensitivities would be possible.

In Figs. 2.6 and 2.7 the statistical behavior of the four designs of Table 2.4 is compared. Noise, linearity, and gain of all the stages are randomly changed in an interval of $\pm 20\%$ from their nominal values. Using MATLAB, half a million samples are made in this way for each design and the NPD of each sample is calculated. The range of NPDs between -72 and -68 dBm is divided into intervals of 0.01 dB and the number of samples falling in each interval is shown in Fig. 2.6. The non-optimum design clearly shows the highest number of out-of-specification

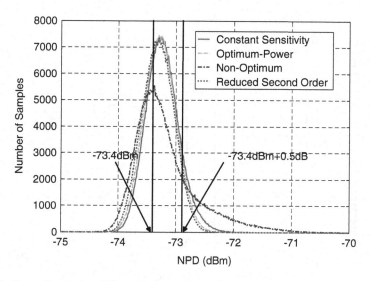

Fig. 2.6 Comparing the four different designs in terms of number of random samples versus NPD

samples, whereas reduced-second-order-sensitivity design shows the lowest number of out-of-specification samples. Figure 2.7 shows the percentage of samples with NPDs smaller than a specific value. For instance percentage of samples with NPDs more than 0.5 dB higher than nominal is 12.15%, 8.3%, 22.45%, and 6.7% for constant sensitivity, optimum-power, non-optimum, and reduced-second-order-sensitivity designs respectively. The percentage of samples with NPDs more than 1 dB higher than nominal is 0.59%, 0.09%, 9.58%, and 0.03% for the aforementioned designs respectively. The reduced second order design achieves 99.9% yield if the acceptable limit for NPD is set to -72.5 dBm, whereas the non-optimum design achieves such yield if that limit is set to -70.7 dBm. If one decides to compensate for this difference by overdesigning the non-optimum case, one has to shift the nominal NPD by 1.8 dB which means that $V_{ni,tot}^2$ must be improved by at least a factor of 1.5 and $1/V_{IP3i,tot}^2$ by at least a factor of 1.22, which in turn increases the power consumption by at least a factor of 1.84 (84%).

The receiver has three stages and each stage has three parameters: noise, linearity, and gain. If each parameter can vary by $\pm20\%$ around its nominal value, a nine-dimensional parameter space is formed with 2^9 corners. Figure 2.8 shows the NPD of each of the four designs of Table 2.4 on the corners of this parameter space.

Apparently some corners yield better performance than the others. However, on the corners which give the worst performance, the non-optimum design has the poorest performance level. It can be seen in Fig. 2.8 that the non-optimum design shows at least 2 dB more degradation of NPD at some corners than the reduced-second-order-sensitivity design which achieves the best corner performance. This means that to achieve the same corner performance from the non-optimum and the

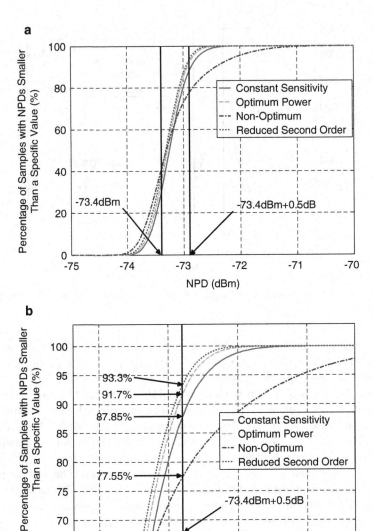

Fig. 2.7 Comparing the four different designs in terms of: (**a**) percentage of random samples with smaller than a specific NPD, (**b**) zoomed view of (**a**)

reduced-second order design, one has to improve the NPD of the non-optimum design by 2 dB which means that $V_{ni,tot}^2$ must be improved by at least a factor of 1.58 and $1/V_{IP3i,tot}^2$ by at least a factor of 1.26, which in turn increases the power consumption by at least a factor of 2 (100%).

As mentioned before, the reduced-second-order-sensitivity approach is more focused on reducing the sensitivity of the NPD to the gains, whereas the constant-sensitivity approach puts constraints on the sensitivity of the NPD to the noise and

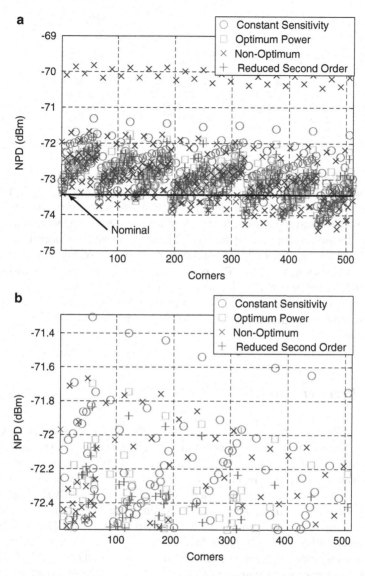

Fig. 2.8 (a) NPDs of designs of Table III on the corners of the parameter space, (b) zoomed view

nonlinearity. Comparing these two approaches in Figs. 2.6, 2.7, and 2.8 reveals that the sensitivity of the NPD to gain variations of individual stages is more influential on the corner and statistical performance than the sensitivity of the NPD to the noise and nonlinearity of individual stages. This is because for a constant total noise and nonlinearity, reducing the sensitivity of the NPD to the noise or nonlinearity of one stage increases the sensitivity to the noise or nonlinearity of other stages, whereas the sensitivity to the gain of all the stages can be reduced simultaneously.

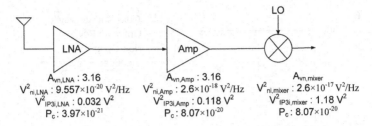

Fig. 2.9 Optimum-power three-stage receiver with different power coefficients

The reason that in the above example the optimum-power design is showing a better corner performance than the constant-sensitivity design is that the NPD has a lower second-order sensitivity to the gain which in turn is caused by the low power coefficient of the last stage (mixer). However if the power coefficients are modified as in Fig. 2.9 the optimum power design will be more sensitive to block-level gains. The NPD histograms for both the optimum-power design of Fig. 2.4 and that of Fig. 2.9 are shown in Fig. 2.10. The percentage of samples with NPDs more than 0.5 dB higher than nominal has increased from 8.3% to 18.57%.

Therefore, simultaneous optimum-robustness and optimum-power can be achieved only if the power coefficients of the rear stages are small compared to those of the front stages.

2.6.2   Receiver with ADC

Now we investigate the possibility of applying the design approaches described in Sect. 2.5 to a 60 GHz receiver including ADC, shown in Fig. 2.11. Considering the performance of state-of-the-art components in the 60 GHz band, the system requirements of a receiver including an ADC must be well more relaxed than a system without an ADC. Due to the inefficiency of traditional ADC/DSP approach at GHz bandwidths, many designers tend to eliminate the ADC from the system and replace it with an analog demodulator or a mixed-signal circuit [18, 19]. In this system-level study, a state-of-the-art ADC reported in Ref. [20] is used. The ADC has a sampling rate of 2.5 GS/s, ENOB of 5.4 bits, VDD of 1.1V, SFDR of -43 dBc, and power consumption of 50 mW. Translating these parameters into the RF domain [21], results in a V_{ni}^2 of 4.53×10^{-14} V^2/Hz and V_{IP3i}^2 of 5. The parameters of the other components are listed in Table 2.5. These parameters are realistic and consistent with state-of-the-art 60 GHz receiver components [16, 17, 20].

Using these components, a total noise figure of 12.9 dB and a total IIP3 of -22.8 dBm can be achieved from the receiver. The total power consumption of the receiver will be 155 mW. Assuming an SNDR of 12 dB and a P_{IMD3i} 3 dB below the noise level, the minimum detectable signal for this receiver would be -54.4 dBm. Assuming an output power of 10 dBm for the desired transmitter and an antenna

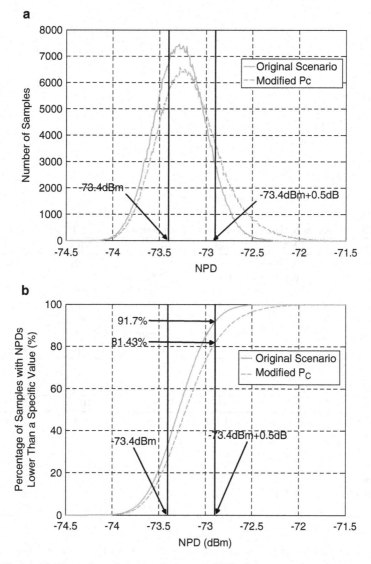

Fig. 2.10 (**a**) Number of random samples versus NPD for two optimum-power designs, (**b**) percentage of random samples with smaller than a specific NPD

gain of 10 dBi for the transmitter and receiver, this MDS associates with a transmission distance of 6.6 m. Considering the total IP3 of the receiver, the maximum tolerable interferer power at the input of the receiver is −38.9 dBm. Assuming an interfering transmitter with an output power of 10 dBm and residing in a 60° direction, making the gain obtained from antenna 7 dBm, the tolerable interference distance will be 56 cm.

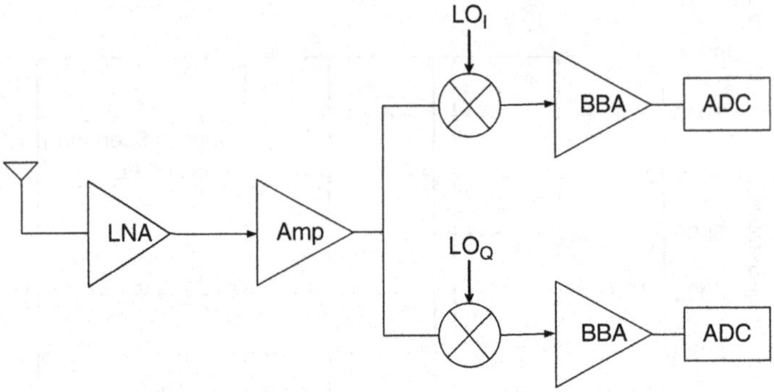

Fig. 2.11 60 GHz zero-IF receiver including ADC

Table 2.5 Component parameters for a non-optimum design including ADC of Ref. [20]

Parameter	LNA	Amp	Mixer	BBA	ADC
V_{ni}^2 ((nV)^2/Hz)	0.2896	1	20	100	4.53×10^4
V_{IP3i}^2 (V^2)	0.1256	0.1256	0.621	1	5
A_{vn}	3.16	3.16	1.3	10	–
NF (dB)	3.8	7.7	19.9	26.8	53.4
IP3 (dBm)	4	4	10.9	13	20
P_C (aW/Hz)	0.0807	0.0807	0.000794	1	906
P (mW)	35	10	0.25	10	100

Table 2.6 Component parameters for an optimum-power design including ADC of Ref. [20]

Parameter	LNA	Amp	Mixer	BBA	ADC
V_{ni}^2 ((nV)^2/Hz)	0.1421	1.421	6.56	55.6	5.38×10^4
V_{IP3i}^2 (V^2)	0.00707	0.0707	1.53	0.52	5.3
A_{vn}	3.16	3.16	1.3	10	–
NF (dB)	2.3	9	15	24.3	54
IP3 (dBm)	−8.5	1.5	14.9	10	20
P_C (aW/Hz)	0.0807	0.0807	0.000794	1	906
P (mW)	4	4	0.19	9.4	89

Doing an optimum-power design according to Table 2.6, requires very-high-performance components which have not been reported in the literature yet. For example the required noise figure of the LNA is 2.3 dB which is well below the best reported figures in the literature [16]. This is due to the fact that the power coefficient of the ADC is much higher than that of the LNA which results in relaxed specifications for the ADC and demanding specifications for the LNA by the optimum-power approach. However the resulting specifications for the LNA are not realistic. Therefore, even with such a relaxed overall requirement, optimum-power design is not yet feasible with state-of-the-art CMOS building blocks. Nevertheless, the total power consumption of a receiver designed based on Table 2.6 would be 106.6 mW which is 31% lower than the receiver in Table 2.5.

Table 2.7 Component parameters for a non-optimum design including ADC of Ref. [22]

Parameter	LNA	Amp	Mixer	BBA	ADC
V_{ni}^2 ((nV)^2/Hz)	0.2896	1	20	100	2.07×10^5
V_{IP3i}^2 (V^2)	0.1256	0.1256	0.621	1	5.7
A_{vn}	3.16	3.16	1.3	10	–
NF (dB)	3.8	7.7	19.9	26.8	60
IP3 (dBm)	4	4	10.9	13	20.6
P_C (aW/Hz)	0.0807	0.0807	0.000794	1	487
P (mW)	35	10	0.25	10	13.4

This unrealistic prediction by the optimum-power method is due to some assumptions made in the course of development of the method which are not always valid. The assumptions are that the noise and nonlinearity performance of an RF circuit can be improved linearly and indefinitely by just increasing the power consumption and that for constant power consumption the noise and linearity performance can be traded for each other, resulting in the prediction of an LNA with 4 mW power consumption and 2.3 dB noise figure. Since these figures are not realistic, we can conclude that these assumptions are not valid in this case or the case is beyond the validity region of the assumptions.

The optimum power design shown in Table 2.6 has a zero first-order sensitivity to the block-level gains, provided that the specified worst case interferer is −38.9 dBm, satisfying the requirement of (2.36). Designing for low second-order sensitivity to block-level gain variations requires the rear stages of the receiver to have lower contribution to the total noise and nonlinearity distortion. However, since in this case the power coefficient of the ADC is much higher than that of the other components, demanding better noise and linearity performance from the ADC dramatically increases the total power consumption of the receiver.

There is another ADC reported in the literature, which, although it does not meet the required sampling rate, has a much better power consumption [22]. With 6.7 mW power consumption, 1GS/s sampling rate, ENOB of 5, and SNDR of 31.5 dB, this ADC can be considered for a 60 GHz receiver with only half the nominal IF bandwidth (500 MHz). The specifications converted to RF domain are shown in Table 2.7. Using the ADC with the same RF components of Table 2.5, results in a receiver with a power consumption of 68.7 mW which is less than half of the power consumption of the receiver of Table 2.5. The resulting total NF and IP3 are 18 dB and −22 dBm, respectively. Although the power consumption is much better, the noise performance is inferior to that of the receiver of Table 2.5. This is not surprising, because, although the power consumption of the ADC of Ref. [22] is 7.5 times smaller than that of Ref. [20], its power coefficient is only 0.54 of that of Ref. [20]. An optimum-power design with the ADC of Ref. [22] results in the specifications given by Table 2.8, which are feasible as compared to the specifications of Table 2.6. Please note that the overall performance parameters are still 18 dB of NF and −22 dBm of IP3 which are relaxing compared to those of Table 2.6.

Table 2.8 Component parameters for an optimum-power design including ADC of Ref. [22]

Parameter	LNA	AMP	Mixer	BBA	ADC
V_{ni}^2 ((nV)^2/Hz)	0.5897	5.897	12.63	231	1.81×10^5
V_{IP3i}^2 (V^2)	0.00682	0.0682	3.181	0.5	6.33
A_{vn}	3.16	3.16	1.3	10	–
NF (dB)	3.85	14.7	17.9	30.5	59.4
IP3 (dBm)	−8.6	1.35	18	10	21
P_C (aW/Hz)	0.0807	0.0807	0.000794	1	487
P (mW)	1	1	0.2	2	17

The field of ADC design for high-speed communication is experiencing a rapid progress and new high-performance ADCs are reported in the literature which may soon make the ADC/DSP approach possible for mm-wave communication links [101].

2.7 Conclusions

Based on a system-level sensitivity analysis performed on a generic RF receiver, it has been shown that the first order sensitivities of the overall performance, represented by NPD, to the individual gains of the blocks can all be made zero. Applying the analysis to a zero-IF three-stage 60 GHz receiver shows a significant improvement in the design yield. A quantity called contribution factor is defined as the noise and nonlinearity distortion contribution of each stage with respect to that of its previous stage. Reduction of the second order sensitivity of the NPD to the gain of individual stages, by keeping the contribution factor of all the stages below one, results in further improvements in the design yield. The conventional optimum-power design methodology has been modified in a way that it nullifies the first order sensitivities of NPD to the individual gains of all the stages. It has been shown that simultaneous optimum-power and optimum-robustness can be achieved by using less power-hungry components at the rear stages of the receiver. Applying the analysis to a 60 GHz receiver including ADC, shows that state-of-the-art ADCs are not adequate for optimum-power or optimum-robustness receiver design at 60 GHz.

Chapter 3
Layout and Measurements at 60 GHz

In this chapter the layout procedure, layout challenges, and measurement setups for 60 GHz circuits designed in this work are explained. Millimeter-wave integrated circuits face additional difficulties in circuit design, layout, and measurement, as compared to their lower frequency counterparts. The parasitic effects are so much accentuated at these frequencies that the designer is required to shift repeatedly the focus from circuit-schematic level to layout and vice versa. In fact the mm-wave circuits may fail to operate correctly just as a result of layout problems. For instance inappropriate interconnect sizing in the layout of a mm-wave voltage-controlled oscillator (VCO) can cause failure of oscillation.

Therefore, in this chapter mm-wave layout challenges and considerations such as the impact of parasitics, substrate losses, dealing with the cross-talk issues, and electromagnetic modeling of complex structures are explained. Then the setups used for measuring the 60 GHz circuits designed in this work are illustrated.

3.1 Layout Challenges

During the preliminary schematic-level design and simulations all the connections between passive and active components are assumed to be realized with perfect lossless and reactance-less wires. However, during the layout the components are connected together via metal layers with associated non-idealities such as losses, capacitances, and inductances. The impact of these non-idealities, referred to as parasitics, turns out to be more significant at higher frequencies. In addition, if the wavelength of the signal is comparable to the dimensions of these interconnects, it is not longer possible to treat them as lumped elements. Therefore, interconnect lines with dimensions comparable to the wavelength of the signal must be modeled with distributed circuits or simulated with electromagnetic simulators. Further-more, the substrate losses, resulting from the currents induced in the substrate,

P. Sakian et al., *RF-Frontend Design for Process-Variation-Tolerant Receivers*,
Analog Circuits and Signal Processing, DOI 10.1007/978-1-4614-2122-1_3,
© Springer Science+Business Media New York 2012

constitute an additional problem which must be considered and solved during the layout. The cross-talk between signal lines and also between components is another issue which is directly related to the positioning of the components in the layout and also to the shielding between the signal lines and components. In the following, these problems along with their corresponding applied solutions are explained in more detail.

3.1.1 Impact of Parasitics

As already mentioned, making a layout comprises connection of passive and active components through metal layers with associated non-idealities such as losses, capacitances, and inductances. The impact of these non-idealities, referred to as parasitics, is more pronounced at higher frequencies. Although the value of capacitances or inductances corresponding to metal connections may not be strongly dependent on the frequency of operation, these parasitic reactive elements become more influential on the performance of the circuits as the frequency of operation is increased. This is mainly due to the fact that in high frequency circuits the required capacitive and inductive components are smaller than those required for low frequency circuits. Therefore, as the frequency of operation is increased, the parasitic reactive elements turn out to be comparable with the required reactive components and hence more difficult to neglect. The tendency of high frequency currents to flow on the surface of conductors, referred to as skin effect, raises the effective resistance of the metal interconnects at higher frequencies. Therefore, the parasitic elements are not limited to capacitances and inductances, and the additional resistances due to the skin effect are part of the non-idealities concerning the layout connections.

The signal wavelength is shorter at higher frequencies. If the wavelength of the signal is comparable to the dimensions of interconnects, they cannot be treated as lumped elements anymore. Therefore, interconnect lines with dimensions comparable to the wavelength of the signal must be implemented by transmission lines with known measured characteristics, modeled with distributed circuits, or simulated with electromagnetic (EM) simulators. The wavelength of a 60 GHz electromagnetic wave in free space is 5 mm. In silicon and silicon dioxide SiO_2, due to the dielectric constant which translates into a reduced propagation speed, the corresponding wavelength of 60 GHz signals is 1.46 and 2.5 mm, respectively. As a rule of thumb any line in the order of one tenth of a wavelength is regarded as distributed.

In this work software tools are used for extracting the resistance and capacitance (RC extraction) of the interconnects. However, these tools do not provide inductance extraction (L extraction), which might be available in some tools not used in this work. Therefore, as a rule of thumb, the metal lines longer than 10 μm, which give rise to considerable inductance values, are manually RLC extracted. Empirical formulae can be used to estimate the capacitance, inductance, and resistance of any metal connection with given dimensions. The magnetic energy storage components such as inductors and transformers which are not supported by accurate circuit models must be simulated in electromagnetic simulation tools.

To address all the above mentioned problems in the design procedure, a methodology must be adopted that takes the parasitics and the distributed nature of long lines into account. Therefore, after finishing the layout the non-overlapping subcircuits which do not include any transmission lines, lines longer than 10 μm, or magnetic energy storage components must be identified and RC-extracted separately. Then the long lines are manually RLC extracted and modeled with distributed circuits. The transmission lines are also modeled with distributed RLC networks. The magnetic storage components are simulated in electromagnetic simulation tools if needed. A new schematic is generated that includes the RC-extracted sub-circuits, the RLC-extracted long lines and transmission lines, and the magnetic energy storage components. Simulating this new schematic can determine if the designed circuit with the given layout meets the required specifications. An example is shown in Fig. 3.1 illustrating the layout of a VCO with four inductors. In Fig. 3.1a the total layout is shown. The lines longer than 10 μm, bondpads, and inductors are removed in the layout of Fig. 3.1b. The remaining long metal lines are carrying DC biasing signals rather than RF signals and hence do not need to be removed for L extraction. After performing RC extraction on the layout of Fig. 3.1b, RLC extraction on the removed lines and replacing the transmission lines, bondpads, and inductors with their models a new circuit schematic is derived as shown in Fig. 3.1c that can be used for further simulations. The long lines are modeled with RLC π-networks after extracting the corresponding values from empirical formulae.

To summarize the adopted methodology, the sequence of tasks in the design procedure is illustrated in the flowchart of Fig. 3.2. First, the required specifications are defined for the circuit. After deciding about the circuit topology and performing the pre-design analyses, the schematic-level simulations are performed. The schematic design is repeated until the specifications are met. Then the layout is done. Afterwards, the layout is split into sub-circuits without parasitic inductances and non-modeled parts and sub-circuits with parasitic inductances (or non-modeled parts). Then, automatic RC-extraction and manual RLC extraction (or modeling) are performed on them, respectively. The new schematic, formed from the RC- and RLC-extracted sub-circuits, is simulated. If the simulations results meet the specifications, the design is finished. Otherwise the layout or schematic-level design must be revised.

At the time of performing the designs of this book, the significance of running an electro-magnetic (EM) simulation on the whole layout was not known to us. Furthermore, running such simulations was found to be very time-consuming even for single components such as inductors when including the patterned shielding. However, now we believe that for an accurate estimation of the performance in the design phase, the whole layout should be simulated in an EM simulator after some simplifications, e.g., replacing the patterned shields by planes. The discrepancies found in this book between the simulation and measurement results can be attributed to the lack of such global EM simulations.

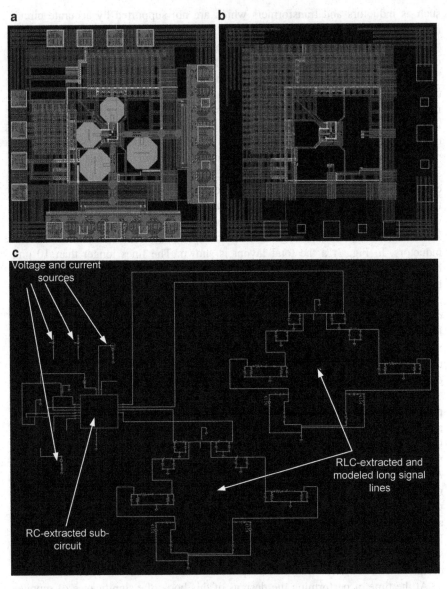

Fig. 3.1 Illustration of the design methodology for including the effect of layout non-idealities: (**a**) the total layout of a VCO, (**b**) the layout prepared for RC extraction, (**c**) new schematic including the RC-extracted sub-circuit and distributed circuit

3.1.2   Electromagnetic Simulation of Complex Structures

As mentioned in the previous section, the un-modeled parts of the layout which involve magnetic energy storage or magnetic coupling must be removed from the layout under RC-extraction and modeled separately. If these un-modeled parts

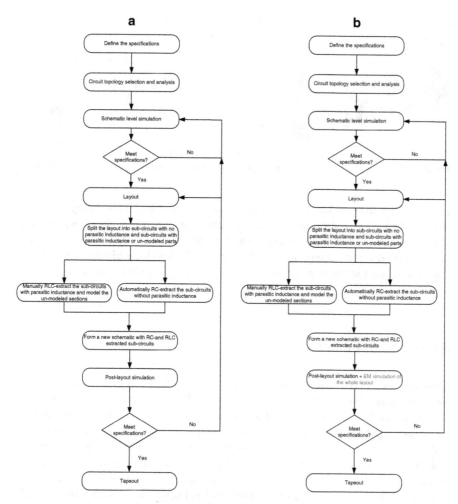

Fig. 3.2 Flow-chart of the: (**a**) adopted design methodology, (**b**) proposed methodology for future designs

consist of simple structures such as regular metal lines they can be modeled manually by empirical formulae. However in case of complex passive structures such as transformers and adjacent inductors with unknown coupling, EM-simulators must be used to find accurate and reliable models for post-layout simulation.

As an example, a 3D view of the layout of a transformer used in a 60 GHz LNA (described in Chap. 4) is shown in Fig. 3.3a. The layout is given to an EM-simulator and the z-parameters of the corresponding two-port are obtained over a wide frequency range. Figure 3.3b shows a circuit model for the transformer. C_1 and C_2 represent the capacitance to substrate and C_M is the interwinding capacitance.

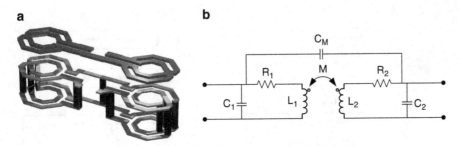

Fig. 3.3 Modeling of a transformer: (**a**) a 3D view of the transformer layout, (**b**) circuit model

L_1 and L_2 are the self-inductance of the primary and secondary windings, respectively, whereas M represents the mutual inductance between them. R_1 and R_2 represent the losses of the windings. These parameters can be extracted from the EM-simulated z-parameters of the transformer. The resulting circuit of Fig. 3.3b can then be used in the new schematic developed for post-layout simulations.

3.1.3 Substrate Losses

Substrate losses comprise another layout-level issue that can degrade the quality factor of the passive components and increase the loss of the transmission lines and interconnects. The low resistivity of the substrate in advanced CMOS technologies, for instance 10 Ω cm in case of bulk 65 nm CMOS, results in formation of undesired electric currents in the substrate as a result of coupling with signal lines and passive components. These currents in combination with the resistance of the substrate give rise to some energy dissipation which reflects itself as a resistance in the circuit model of the transmission lines, inductors, and other signal lines and passive components. As a result of this energy dissipation in the substrate, the quality factor (Q) of inductors is degraded and the loss of transmission lines is increased. High-Q inductors are necessary for implementing high-gain RF amplifiers and high-performance voltage controlled oscillators. Low-loss transmission lines are required for realizing loss-less matching networks.

Some approaches have been proposed to deal with the problem of the substrate loss. For instance using high-resistivity (150–200 Ω cm) silicon substrate is proposed to imitate the low-loss semi-insulating GaAs substrate [23]. However this is not a commonly available option in regular silicon technologies. Utilizing silicon-on-insulator (SOI) technologies with 1 kΩ cm substrate resistivity is another solution which can provide high-Q passive components and low-loss transmission lines [24]. But, the additional cost due to the required extra lithography steps is the main drawback of this option. Another proposed solution is etching a pit in the silicon substrate under the inductors and transmission lines to remove the substrate

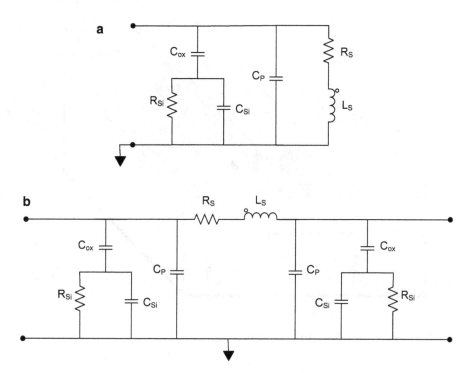

Fig. 3.4 Lumped physical models for: (**a**) a spiral inductor on silicon substrate, (**b**) a transmission line on silicon substrate

effects [25]. However, the etch results in additional processing cost and causes reliability concerns such as long-term mechanical stability.

A proposed low-cost solution to the substrate loss problem involves shielding the electric field of the components from the silicon substrate [26]. To explain this method, the circuit models of an inductor and a transmission line including the substrate effects are illustrated in Fig. 3.4a, b, respectively [27, 28]. Although an on-chip inductor is physically a four-port component including the substrate and the shield, the one-port connection of Fig. 3.4a captures the inductor characteristics while avoiding unnecessary complexities. L_S represents the spiral inductance in Fig. 3.4a and the series inductance per unit length in Fig. 3.4b. R_S is the series resistance which accounts for the metal resistance, the skin effect, and the energy losses due to the eddy current induced in conductive media close to the inductor or transmission line. Such conductive media include any metal layer in the vicinity of the inductor or transmission line and also the substrate. Therefore, part of the substrate losses are caused by the eddy currents induced in the substrate as a result of the time-varying magnetic fields around inductors or transmission lines which penetrate into the substrate. The rest of the substrate losses are accounted for by R_{Si} which stands for the silicon substrate resistance. C_{Si} stands for the silicon substrate capacitance. C_{ox} is the oxide capacitance between the metal layer and

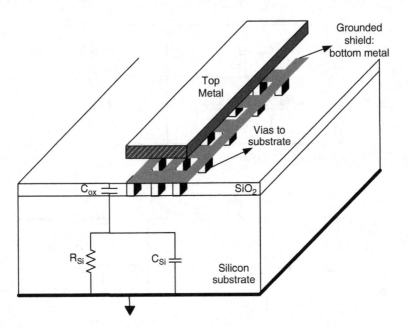

Fig. 3.5 Grounded patterned shield with substrate contacts

the substrate. C_P accounts for the capacitances between the signal-carrying metal layer and the ground, for instance, in case of the spiral inductor it includes the capacitance between the spiral and the center-tap underpass [29]. Any penetration of the electric field into the substrate generates currents in the substrate resistance (R_{Si}) and hence results in energy dissipation. It is revealed from Fig. 3.4 that the substrate losses can be reduced by either increasing R_{Si} to infinity or reducing it to zero.

Therefore, it is possible to improve the quality factor of the inductor or reduce the loss of the transmission line by making the silicon substrate short or open. The aforementioned costly methods of using a high-resistivity substrate or etching a pit in the substrate are examples of attempting to provide an open-circuit substrate.

Alternatively, using grounded shielding metal layers with substrate contacts can provide a short-circuit substrate with lowered resistive losses, as shown in Fig. 3.5. This can be explained physically by regarding the grounded metal layers as shielding planes which prevent the electric field from penetrating into the substrate. Equivalently it can be explained by the circuits of Fig. 3.4 showing that a short-circuited R_{Si} carries no current and hence makes no contribution to the losses. However, the eddy currents induced in the shielding metal layers reduce the effective value of the inductance, increase the parasitic capacitance of the transmission line (C_P), and add to the losses.

To address this problem, patterned ground shields are used to avoid eddy currents while preserving the required low resistance shunt with the substrate [26]. It is also possible to use floating shield in a virtual ground position for differential

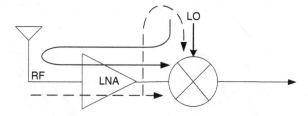

Fig. 3.6 Coupling between the LO and RF paths in a receiver

transmission lines to exploit the advantages of ground shields while avoiding a physical connection between the shield and the substrate [30].

In this work patterned ground shields are available and their models are verified by measurements. Therefore, this type of shielding is utilized in the layout of the inductors and transmission lines, as shown in Fig. 3.5, to reduce the substrate-induced losses. Poly and Metal 1 layers are used to keep the maximum distance with the signal lines and avoid increasing parasitic capacitances from the signal lines to the ground.

3.1.4 Cross-Talk Shielding and Grounding

Cross-talk and coupling between different signal lines and passive components is a source of several problems in RF circuits and systems. Noise injection from DC supplies or the substrate can deteriorate the performance of circuits such as voltage-controlled oscillators or low-noise amplifiers. Coupling between the RF and LO signal paths in a zero-IF receiver, as shown in Fig. 3.6, leads to DC offset and second-order intermodulation distortions.

There are at least four different sources of cross-talk in integrated RF circuits, including the substrate coupling, magnetic mutual coupling between inductors, magnetic and capacitive coupling between signal lines, and imperfect ground and DC lines.

The problem of substrate coupling can be alleviated by shielding and grounding the substrate. Therefore, the same approaches used for resolving substrate loss issues can be useful for overcoming the substrate coupling problem as well. Since inductors and transmission lines usually occupy substantial chip area, they can be both sources and receptors of noise coupling. Therefore, shielding them from the substrate in a way that their electric field is terminated before reaching the substrate can significantly reduce the cross-talk via the substrate. Using substrate contacts to connect the substrate to grounded metal layers can be helpful in suppressing the undesired RF signals coming into the substrate from different sources. Therefore, it is essential to form rings of grounded substrate contacts around critical RF blocks such as transistors or passive components.

The undesired magnetic coupling between inductors and the coupling between different signal lines can be avoided by proper positioning and keeping the required

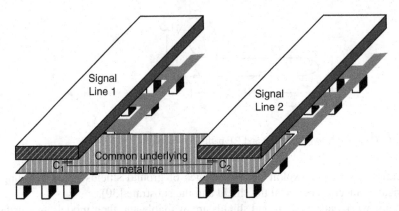

Fig. 3.7 An underlying metal layer forming capacitances with two different signal lines and contributing to their coupling capacitance

Fig. 3.8 Distribution of ground and DC supply lines in an empty area in the vicinity of a bondpad

distance. Furthermore, metal layers passing from below or above different signal lines should be avoided (illustrated in Fig. 3.7), because they contribute to unwanted capacitive coupling between unrelated signal lines.

Imperfect ground or DC lines result in the leakage of RF signals generated in one part of the chip to the other parts. In other words, the DC lines, ideally supposed to be perfect RF-grounds, can transform into RF-signal-carrying lines if the necessary measures are not taken during the layout. To address this potential problem ground lines are distributed in all the empty areas of the chip to provide a low-impedance nearly-perfect ground reference. All the DC supply lines carrying substantial amounts of current are treated in the same way as the ground lines in the layout. All DC lines are connected via big decoupling capacitors to the ground. The DC supply lines which carry almost no current, like the biasing voltage of the gate of transistors and varactors, are not distributed but are decoupled via big capacitors in the vicinity of the circuit. Figure 3.8 shows an example of ground and DC supply distribution, whereas Fig. 3.9 shows decoupling capacitances used for decoupling DC biasing lines.

Decoupling capacitors

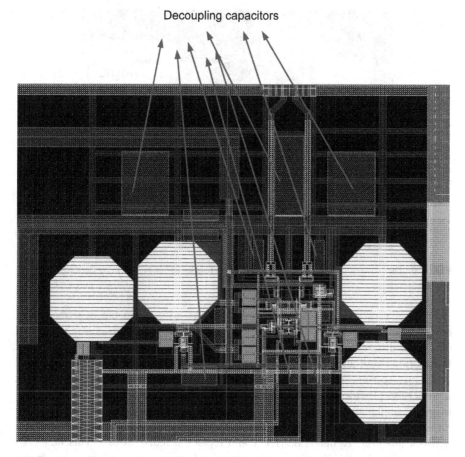

Fig. 3.9 Decoupling capacitors used, connecting the DC supply lines to the ground

3.2 Measurement Setup

All the circuits designed in this work are measured using on-wafer probing. Although this method provides a quick way of testing the circuits, it also raises some difficulties particularly at mm-wave frequencies. The very short wave-length of the signals at mm-wave frequencies makes the measurements very sensitive to the effective length and bending of the interfaces. Especially to perform on-wafer measurements one must pay utmost attention to the rigidity of the interfaces connected to the probes to keep all the connection lengths and orientations constant during the whole period of the measurement and calibration. Also special care must be taken to preserve the position of the probes on the bondpads and impedance standard substrates, since the measurement accuracy can be very much dependent on the positioning and landing of the probes.

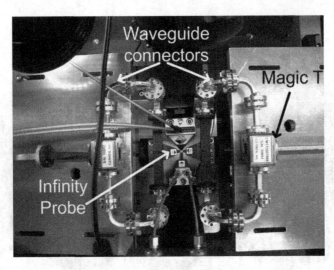

Fig. 3.10 The waveguide-based setup including two magic-Ts for measuring a double-balanced mixer

Another difficulty of mm-wave measurements arises from the overwhelming cost of equipments needed for instrumentation. For instance, s-parameter measurement of a differential two-port mm-wave circuit would require a very expensive four-port network analyzer.

In this section, measurement setups are introduced which use waveguide interfaces to provide the required rigidity in the vicinity of the probes and utilize magic-T single-ended-to-differential converters to facilitate the measurement of differential circuits. As a result of using the magic-T baluns, a two-port power network analyzer, Agilent E8361A, could replace a four-port one. This network analyzer provides 94 dB of dynamic range and covers a frequency range from 10 MHz to 67 GHz. The waveguide setup used for on-wafer measurement of the differential circuits is illustrated in Fig. 3.10. It shows a double-balanced mixer as the device under test (DUT) which requires four probes for measurement. The probe on the top of the picture is an eye-pass probe used for biasing. The probe at the bottom of the picture is a Cascade GSGSG microprobe suitable for up to 50 GHz measurements and used here at the IF output of the DUT mixer. The other two probes on the left and right side are Cascade infinity GSGSG probes suitable for mm-wave signals and used here at the RF and LO differential inputs of the mixer. The waveguide structures are mounted on metal plates which are screwed to the probe station, preventing all kinds of unintentional movements in the setup.

Since there was no 60 GHz signal generator yet available in the measurement laboratory, the network analyzer was sometimes used in signal-generation mode to provide the input to 60 GHz circuits. When more 60 GHz signals were needed, a two-box solution illustrated in Fig. 3.11 was used. It combines a signal generator which generates signals up to 40 GHz with an upconverter provided by another signal

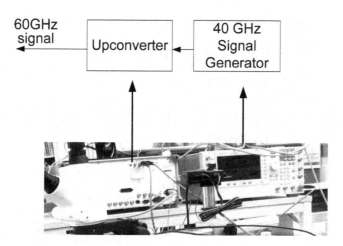

Fig. 3.11 Using a 40 GHz signal generator in combination with the upconverter of another signal generator to produce a 60 GHz signal

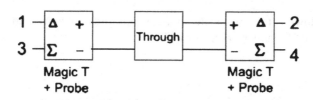

Fig. 3.12 Two magic-Ts cascaded via probes and a through to measure the differential-to-differential and common-mode-to-differential transfer functions

generator to generate a 60 GHz signal. An advantage of this solution compared to using the network analyzer is the possibility of generating two-tone signals.

Waveguide-based magic-Ts are extensively used in this work for measuring differential circuits. Therefore, it is important to know the quality of their single-ended-to-differential conversion. Figure 3.12 shows a cascade of two magic-Ts used for measuring the isolation between differential and common-mode. Figure 3.13 compares the measured common-mode-to-differential leakage with the differential-to-differential loss of such structure, showing at least 20 dB difference between the differential signal and the undesired common-mode.

Measuring 60 GHz signals was not straightforward either, as there were no spectrum analyzers and signal analyzers available yet in the measurement laboratory. Agilent E4446A series spectrum analyzer covers the frequency range of 3 Hz–44 GHz. It is used in combination with Agilent 11,974 V preselected V-band mixer to allow signal measurement at 60 GHz. Alternatively, Agilent E5052A signal source analyzer can be used in combination with a passive downconverting mixer to enable signal analysis at 60 GHz.

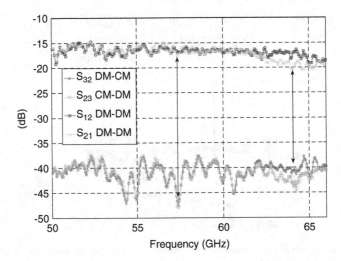

Fig. 3.13 Measured s-parameters of the four-port of Fig. 3.12: showing at least 20 dB difference between common-mode-to-differential and differential-to-differential

In the following, the s-parameter, noise, nonlinearity, and phase noise measurement setups used in this work are described.

3.2.1   Calibration and s-Parameter Measurement

Performing s-parameter measurements on differential circuits with a two-port network analyzer is facilitated by utilizing the magic-Ts. As shown in Fig. 3.14, each port of the network analyzer is connected to a magic-T and then to the probes. SOLT (Short-Open-Load-Through) calibrations are performed on a standard impedance substrate, suitable for GSGSG probes. Then the impedance standard substrate is replaced by the DUT and the measurement is done.

To measure the common mode response of a differential circuit the magic-Ts can be connected in such a way that they provide common mode (CM) signals at input and output.

Conforming to the following considerations can promote the accuracy of the measurements and calibrations:

1. Accurate definition of the impedance standard substrate in the network analyzer or the software which controls the network analyzer
2. Precise positioning of the probes on the bondpads or on the impedance standard substrate
3. Repeating the calibration after some time due to invalidity of the calibration results after a certain period
4. Using undamaged samples of impedance standard substrate

Fig. 3.14 s-parameter
measurement and calibration
setup of a differential two-
port circuit

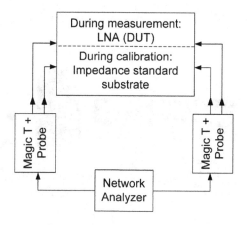

3.2.2   Noise Measurement

In this section, the setups used for the noise measurement of a double-balanced
60 GHz mixer and a 60 GHz differential LNA are described.

Figure 3.15 shows the block diagram of the setup used for the noise figure
measurement of a 60 GHz double-balanced mixer using Y-factor method [31].
The network analyzer is used to produce the 60 GHz LO signal, as explained before.
The 60 GHz noise source is connected via an isolator (to ensure 50 Ω termination
for the noise source) and a waveguide to the magic-T and then to the RF port of the
mixer. The differential IF output of the mixer is converted to single-ended via a
hybrid and then connected to the spectrum analyzer via a low-frequency amplifier
which covers 30 MHz–4 GHz. The spectrum analyzer is set to noise figure mode
with the DUT specified as a downconverter with a 60 GHz LO. The IF frequency
range is set to 30 MHz–2 GHz. The 60 GHz noise source is characterized only in the
range of 60–75 GHz. Therefore, another noise source is needed for calibration of
the output path and the spectrum analyzer. Figure 3.16 shows the block diagram of
the noise calibration setup. The low-frequency amplifier is essential for obtaining
good calibration results by boosting the noise above the noise floor of the spectrum
analyzer. Sometimes even more amplifiers are needed for this purpose. Since two
different noise sources are used, the ENR (excess noise ratio) list of the two noise
sources must be manually entered in the spectrum analyzer. Both noise sources are
controlled by the spectrum analyzer.

The effect of the low-frequency amplifier and the cable, connecting the IF balun
to the low-frequency amplifier, is automatically taken into account during the
measurement, because they are in the calibration setup. However, the impact of
the IF balun and the RF interfaces at the input of the DUT must be manually
calculated after the measurement. The loss of the combination of the magic-T,
waveguide structure, and the infinity probe is measured via two methods for
confirmation. The first method is illustrated in Fig. 3.17. Using a delta measurement

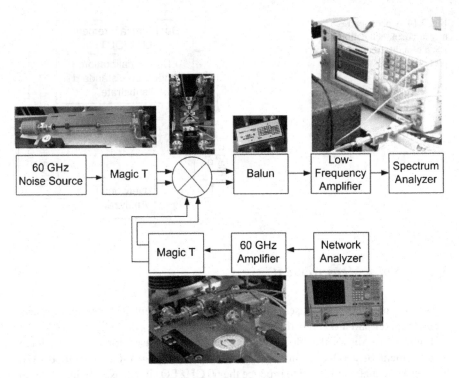

Fig. 3.15 Noise figure measurement setup for the mixer

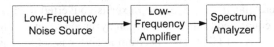

Fig. 3.16 Noise figure calibration setup for the mixer

and utilizing the network analyzer as a signal generator, the amplitude of the 60 GHz signal is measured by the spectrum analyzer, as shown in Fig. 3.17a. Since the available spectrum analyzer does not support 60 GHz range, the Agilent preselect millimeter mixer is used to downconvert the 60 GHz signal to the supported range of the spectrum analyzer. Keeping the source signal amplitude constant, the magic-T and the probes are introduced into the setup, as shown in Fig. 3.17b. A through of an impedance standard substrate is used between the probes. The difference in the readings of the two steps gives the loss of the introduced interface. Assuming a negligible loss for the through and equal loss for the two probes and magic-Ts, the loss of the RF interface, used between the noise source and the mixer input, can be calculated by dividing this number by two.

In the second method, two one-port calibrations are performed using the network analyzer. First a cable, used in the next step for connecting the network analyzer to the magic-T and probe, is calibrated and the calibration dataset is saved. Then an

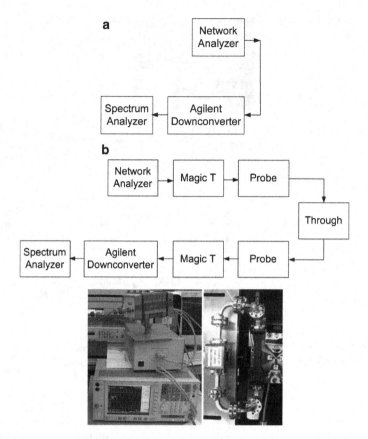

Fig. 3.17 Using delta measurement to measure the loss of the interface between the noise source and the mixer input: (**a**) measuring without the RF interface under test, (**b**) measuring with the RF interface under test

on-wafer one-port calibration is performed using an impedance standard substrate and including the magic-T and the probe in the setup. Again the calibration dataset is saved. Having the two datasets, the s-parameters of the combination of the magic-T and the probe is calculated. The measured loss is the same as the one obtained from the first method (delta measurement).

The noise figure measurement of the 60 GHz LNA is impeded by the fact that the output of the LNA is at a higher frequency than supported by the spectrum analyzer. Even the Agilent preselect mixer cannot be used in this case because the noise figure mode of the spectrum analyzer does not support it and it cannot be used with an external LO either. Therefore a passive mm-wave mixer is used in the noise measurement setup, as shown in Fig. 3.18, to downconvert the output of the LNA to the range of the spectrum analyzer. The passive mixer can be included in the calibration setup as shown in Fig. 3.19, obviating the need for its inclusion in post-measurement calculations.

Fig. 3.18 LNA noise figure
measurement setup

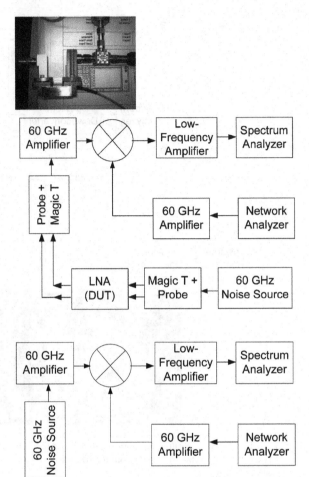

Fig. 3.19 Noise figure
calibration for the LNA

3.2.2.1 Noise Measurement Uncertainty Due to Temperature

To calculate the impact of the lossy interfaces on the noise figure measurement one
has to take the environment temperature into calculation. In fact the noise factor of
a lossy passive network at any temperature different from $T_0 = 290$ K cannot be
obtained by just inverting the loss and is calculated from below [32, 33]:

$$NF = (L - 1)\frac{T}{T_0} + 1 \qquad (3.1)$$

where L is the loss of the network and T is the physical temperature of the network.
In this work this effect is not taken into account, leading to an uncertainty in the
results. However, since the expected temperature deviations from T_0 are in
the order of 4%, the resulting error is not deemed significant. Figure 3.20 shows

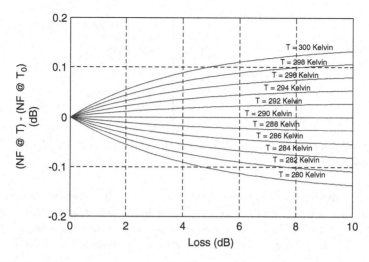

Fig. 3.20 The difference between the noise figures of a lossy passive network at T_0 and at other temperatures as a function of its loss

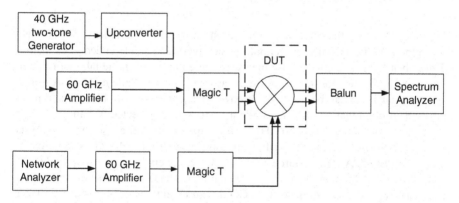

Fig. 3.21 Two-tone test setup for nonlinearity characterization of a double balanced mixer

the deviation of the noise figure of a lossy passive network from its value at T_0 as a function of its loss and at different temperatures. Even at 300 K and for a loss of 10 dB the noise figure is barely 0.15 dB higher than that at 290 K.

3.2.3 Nonlinearity Measurement

In this work the nonlinearity of LNAs and mixers are characterized by two-tone or multi-tone tests. Figure 3.21 shows an example of a two-tone nonlinearity characterization for the double balanced mixer described in Chap. 4. The two-box solution,

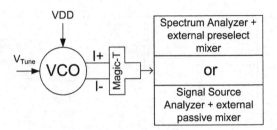

Fig. 3.22 Phase noise measurement setups at 60 GHz: using either spectrum analyzer or signal source analyzer

consisting of the 40 GHz signal generator and the upconverter, is used to generate a two tone signal at 60 GHz. Since the upconverter is highly nonlinear, the level of the signal coming from the 40 GHz signal generator is kept low, while using a linear 60 GHz amplifier to provide enough amplitude for the two-tone signal fed to the DUT.

3.2.4 Phase Noise Measurement

Figure 3.22 shows the setup used for the phase noise measurement of a 60 GHz differential VCO. The differential output is converted to single-ended via a magic-T and then fed to the phase-noise measurement equipment. In the laboratory there are two candidates for phase noise measurement: An Agilent E4446A series spectrum analyzer can be used which covers frequencies up to 44 GHz and must be used in combination with an Agilent 11,974 V preselected V-band mixer to allow noise measurements at 60 GHz. This equipment uses direct spectrum analysis method and is not suitable for drifting free-running VCOs [34, 35]. Alternatively, an Agilent E5052A signal source analyzer, which covers frequencies up to 7 GHz, can be used in combination with an external mixer. The latter has the advantage of a more specialized user interface for VCO measurements. However, since E5052A uses (a cross-correlation implementation of) PLL method for phase noise measurement, it can bias the data close to the carrier as a result of the high loop gain of the PLL [36, 37]. That may occur when the DUT is a drifty free-running VCO which demands a high loop gain from the PLL to stabilize its oscillation frequency [38]. As a result, E5052A must be used with care and the measurement data affected by the PLL must be corrected or ignored.

3.3 Conclusions

The layout challenges at mm-wave frequency range are discussed in this chapter. The parasitic effects due to layout, which are more influential at high frequencies, are taken into account by performing automatic RC extraction and manual L extraction. The long signal lines are modeled with distributed RLC networks. The problem of substrate losses is addressed by using patterned ground shields in inductors and transmission lines. The cross-talk issue is treated by using distributed meshed ground lines, decoupled DC lines, and grounded substrate contacts around sensitive RF components.

The on-wafer measurements on the 60 GHz circuits designed in this work are performed using a waveguide-based measurement setup. The fixed waveguide structures, specially provided for the probe station, serve for the robustness of the setup as they circumvent the need for cables, which are by nature difficult to rigidify, in the vicinity of the probes. Taking advantage of magic-Ts, it is possible to measure differential mm-wave circuits with a two-port network analyzer rather than using a much more expensive four-port one. Also the CM response of differential circuits can be measured by reconfiguring the magic-T to provide CM excitation. Noise, s-parameter, and phase noise measurements are performed using the mentioned setups.

Chapter 4
Component Design at 60 GHz

In this chapter several 60 GHz components are presented and designed in standard CMOS technologies with intrinsically high performance without exhibiting smartness for post-fabrication performance fine tuning. In Chap. 5, smart component will be discussed which are capable of performance tuning for process spreading compensation.

The design, fabrication, and measurement of a low-noise amplifier (LNA), a zero-IF mixer, and a quadrature voltage-controlled oscillator (VCO) are described in Sects. 4.2, 4.3, and 4.4, respectively.

4.1 Low Noise Amplifier

As the first stage of the receiver frontend after the signal source or the antenna, the LNA is responsible for amplification of the signal from the antenna with minimum added noise and distortion. Therefore, noise figure, gain, and linearity are among the main design concerns. Furthermore, the LNA must be stable to prevent self-oscillation. In this section, different LNA topologies are compared in terms of performance and suitability for mm-wave design. Then a theoretical analysis is presented for the selected topology. Finally the circuit design and experimental results are presented.

4.1.1 Topology Selection

One of the most commonly used LNA topologies is the inductively-degenerated common-source LNA, shown in Fig. 4.1 [39]. The inductors L_S and L_G provide the required input matching. The inductor L_S produces the real part of the input

P. Sakian et al., *RF-Frontend Design for Process-Variation-Tolerant Receivers*,
Analog Circuits and Signal Processing, DOI 10.1007/978-1-4614-2122-1_4,
© Springer Science+Business Media New York 2012

Fig. 4.1 An inductively
degenerated common-source
LNA

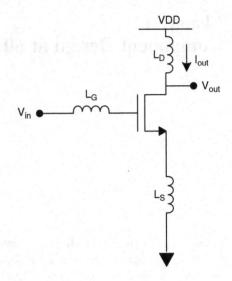

impedance and L_G nullifies the imaginary part. One of the main advantages of this
topology is the implementation of the real part of the input impedance without
having to use a lossy resistor in the signal path. The input-referred noise voltage of
the LNA and its input impedance are described by (4.1) and (4.2), respectively:

$$\bar{v}_{ni}^2 = \frac{4KT\gamma g_{d0}}{g_m^2}$$

$$\times \frac{\left(R_S^2 + \left((L_G + L_S)\omega - \frac{1}{c_{gs}\omega}\right)^2\right)\left(\frac{g_m^2 L_S^2}{c_{gs}^2} + \left((L_G + L_S)\omega - \frac{1}{c_{gs}\omega}\right)^2\right)c_{gs}^2\omega^2}{\left((L_G + L_S)\omega - \frac{1}{c_{gs}\omega}\right)^2 + \left(\frac{g_m L_S}{c_{gs}} + R_S\right)^2}$$

$$\text{(4.1)}$$

$$Z_{in} = \left(\frac{g_m L_S}{c_{gs}}\right) + j\left((L_G + L_S)\omega - \frac{1}{c_{gs}\omega}\right) \qquad \text{(4.2)}$$

where K is the Boltzmann constant, T is the temperature, γ is the channel noise
coefficient, g_{d0} is the channel conductance when the drain-source voltage is zero,
R_S is the source resistance, g_m is the transconductance of the transistor, c_{gs} is the
gate-source capacitance, and ω is the angular frequency. In a good design, L_S
is adjusted in such a way that the real part of Z_{in} is matched with R_S, and L_G is
selected in a way that the imaginary part of Z_{in} is nullified. The multiple influential
design parameters provide a desirable decoupling between different performance
requirements. For instance, the real part of the input impedance can be adjusted
independently while achieving the required noise performance.

  In order to investigate the stability, a simplifying assumption which is made in
the above calculations must be discarded; the gate-drain capacitance, c_{gd}, must be

Fig. 4.2 A voltage–voltage transformer feedback LNA

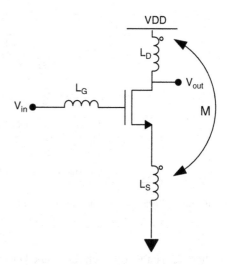

taken into consideration. This capacitance constitutes a reverse path between the input and output which endangers the stability. Several solutions have been proposed to overcome this problem. Using a cascode architecture is one way to implement isolation between the output and input [40]. However at high frequencies the additional parasitics due to the cascode device deteriorate the performance. This has motivated the designers to adopt more complicated cascode architectures using additional inductors [41]. Using a common-gate topology is another way to provide the required isolation. However the common-gate LNAs suffer from a tradeoff between the input matching and the noise figure [42–44]. In this work the isolation is implemented by equipping L_D and L_S with mutual magnetic coupling [16, 45]. The resulting topology is called voltage–voltage transformer feedback LNA and is shown in Fig. 4.2.

4.1.2   Stability and Noise Analysis

Stability is one of the main concerns in radio frequency amplifier design. Any reverse signal path from the output to the input can make the amplifier potentially instable. Therefore, unilateralization, or providing reverse isolation between the output and the input, is a widely used technique to ensure the stability of amplifiers. For every linear two-port device a property is defined which is invariant with respect to certain types of transformations [46, 47] implemented by linear lossless reciprocal embedding four-ports, shown in Fig. 4.3:

$$U = \frac{|\det[Z - Z_t]|}{\det[Z + Z^*]} = \frac{|z_{12} - z_{21}|^2}{4\mathrm{Re}[z_{11}] \cdot \mathrm{Re}[z_{22}] - \mathrm{Re}[z_{12}] \cdot \mathrm{Re}[z_{21}]} \tag{4.3}$$

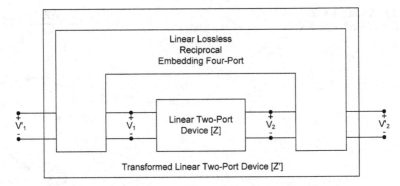

Fig. 4.3 A linear two-port device embedded within a linear lossless reciprocal four-port

where Z is the open-circuit impedance matrix of the two-port, t denotes the transposition and * denotes the complex conjugate. Interestingly this invariant property predicts the maximum gain after unilateralization; i.e., if the original linear two-port is unilateralized by the embedding network, the value of the maximum gain can be obtained from U [47]. Since U is invariant to the transformation done by the linear lossless reciprocal embedding network, the maximum gain can also be calculated by the z-parameters of the unilateralized device ($z'_{12} = 0$):

$$U = \frac{|z'_{21}|^2}{4\mathrm{Re}[z'_{11}] \cdot \mathrm{Re}[z'_{22}]} \qquad (4.4)$$

It is worth noting that U is not the maximum gain obtainable from the two-port with an arbitrary four-port embedding. In fact the maximum obtainable gain is infinite, for instance when the two-port is used in an oscillator circuit. Therefore it is important to note that U is the maximum gain only when the two-port is unilateralized; i.e., when there is no feedback between the input and output. If the condition of unilateralization is changed to stability, it can be shown that the maximum stable gain of a two-port can be as large as 4U for large values of U [48].

The LNA of Fig. 4.2 is visualized in Fig. 4.4 by an analogy with the embedding scenario of Fig. 4.3. The transistor is regarded as a linear two-port device, whereas the coupled inductors, or the transformers, form the embedding four-port. The z-parameters of the transistor are calculated below:

$$Z = \begin{bmatrix} z_{11} & z_{12} \\ z_{21} & z_{22} \end{bmatrix} = \frac{1}{g_m + c_{gs}s} \times \begin{bmatrix} 1 & 1 \\ 1 - \frac{g_m}{c_{gd}s} & 1 + \frac{c_{gs}}{c_{gd}} \end{bmatrix} \qquad (4.5)$$

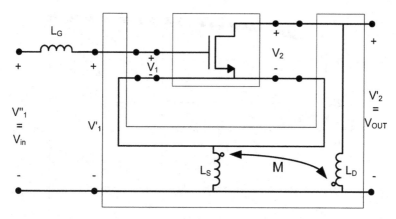

Fig. 4.4 The LNA of Fig. 4.2 illustrated as a two-port device (a single transistor) embedded in a linear reciprocal four-port (transformer)

After embedding, the z-parameters of the transformed network can be obtained from below:

$$Z' = \begin{bmatrix} z'_{11} & z'_{12} \\ z'_{21} & z'_{22} \end{bmatrix} \tag{4.6}$$

where z'_{11}, z'_{12}, z'_{21}, and z'_{22} are calculated from (4.7), (4.8), (4.9), and (4.10). The effect of L_G will then be just a series reactance added to z'_{11}.

$$z'_{11} = z_{11} + L_S s - (z_{12} + L_S s + Ms) \times \frac{z_{21} + L_S s + Ms}{z_{22} + L_S s + L_D s + 2Ms} \tag{4.7}$$

$$z'_{12} = -Ms + (L_D s + Ms) \times \frac{z_{12} + L_S s + Ms}{z_{22} + L_S s + L_D s + 2Ms} \tag{4.8}$$

$$z'_{21} = z_{21} + L_S s - (z_{22} + L_S s + Ms) \times \frac{z_{21} + L_S s + Ms}{z_{22} + L_S s + L_D s + 2Ms} \tag{4.9}$$

$$z'_{22} = -Ms + (L_D s + Ms) \times \frac{z_{22} + L_S s + Ms}{z_{22} + L_S s + L_D s + 2Ms} \tag{4.10}$$

Defining the coupling factor, k_c, and the turn ratio, n, in (4.11) and (4.12), z'_{12} is converted to (4.13), after substitution of z-parameters from (4.4) in (4.8) and some manipulation. From (4.13) it is clear that z'_{12} can be made zero if the conditions of (4.14) are satisfied. By making z'_{12} zero, the required unilateralization is achieved.

$$k_c = \frac{M}{\sqrt{L_D L_S}} \tag{4.11}$$

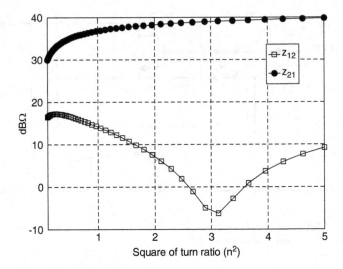

Fig. 4.5 z''_{21} and z''_{12} of the amplifier versus the turn ratio of the transformer showing the minimization of z''_{21} by proper choice of turn ratio

$$n = \sqrt{\frac{L_D}{L_S}} \tag{4.12}$$

$$z'_{12} = \frac{L_D s \left(c_{gd} - \frac{k_c}{n} \times c_{gs}\right) + \frac{1-k_c^2}{n^2} \times c_{gd} L_D^2 s^2 \left(g_m + c_{gs} s\right)}{\left(c_{gs} + c_{gd}\right)s + \frac{c_{gd} L_D}{n^2} \left(1 + 2k_c n + n^2\right)s^2 \left(g_m + c_{gs} s\right)} \tag{4.13}$$

$$\begin{cases} k_c = 1 \\ n = \frac{c_{gs}}{c_{gd}} \end{cases} \tag{4.14}$$

In case of unilateralization which is obtained by meeting the conditions of (4.14), the input impedance of the LNA of Figs. 4.2 and 4.4 is derived from (4.15) which shows that the real part of the input impedance is zero (in fact limited to the gate resistance of the transistor) and the imaginary part can be made zero by properly adjusting the value of L_G.

$$z_{in} = L_G j\omega + z'_{11} = L_G j\omega + \frac{1}{j\omega\left(c_{gs} + c_{gd}\right)} \tag{4.15}$$

Figure 4.5 shows the simulated z''_{21} and z''_{12} as a function of the turn ratio (n). It is seen that z''_{12} is minimized at around 3.1 whereas z''_{21} increases monotonically with n^2. The simulation is done by SPECTRE-RF periodic-scattering-parameter analysis at 60 GHz. The transistor width and transistor length are 32 μm and 65 nm, respectively. The biasing voltage of the gate of the transistor and V_{DD} are 0.7 and 1.2 V, respectively. L_S is kept constant at 500 pH while n^2 and hence L_D are swept from 0.1 to 5 and from 5 pH to 12.5 nH, respectively. L_G is 207 pH to nullify the imaginary part of the input impedance.

In case of unilateralization, if R_G (the gate resistance of the transistor) is not neglected and when L_G is in resonance with the capacitive part of z_{in} and z'_{11}, the z parameters of the circuits of Fig. 4.4 are derived from the following:

$$z''_{11} = R_G \tag{4.16}$$

$$z''_{12} = z'_{12} = 0 \tag{4.17}$$

$$z''_{21} = z'_{21} = \frac{-g_m L_D}{1 - c_{gd} L_D (c_{gs} + c_{gd}) \omega^2 + \frac{g_m c_{gd} L_D}{c_{gs}} (c_{gs} + c_{gd}) j\omega} \tag{4.18}$$

$$z''_{22} = z'_{22} = \frac{L_D c_{gs} j\omega}{1 - c_{gd} L_D (c_{gs} + c_{gd}) \omega^2 + \frac{g_m c_{gd} L_D}{c_{gs}} (c_{gs} + c_{gd}) j\omega} \tag{4.19}$$

The value of U can be calculated in two ways: the z' parameters from (4.16), (4.17), (4.18), and (4.19) can be substituted in (4.3) or (4.4); or the z parameters from (4.5) can be substituted in (4.3) after adding R_G to z_{11} to include the transistor gate series resistance. As expected and as evidence to the invariability of U under the transformation of Fig. 4.4, both approaches yield the same result:

$$U = \frac{g_m}{4 R_G c_{gd} \omega^2 (c_{gs} + c_{gd})} \tag{4.20}$$

In order to perform a noise analysis, the circuit schematic of the LNA including the transistor channel noise and the source noise is shown in Fig. 4.6a. Considering the transistor as a linear two-port, the channel noise can be transferred back to its gate, as shown in Fig. 4.6b, using the T-parameters of the transistor. Then the resultant noise current source between the gate and the source of the transistor can be split into two, as shown in Fig. 4.6c. Considering the whole circuit as a two-port, with one port between the gate of the transistor and the ground and the other port between the source of the transistor and the ground, and calculating the T-parameters of this two-port, the noise current source in parallel with L_S can be transferred back to the input, as shown in Fig. 4.6d.

According to Fig. 4.6, only the B and D elements of the T matrices are needed. Therefore the B and D elements of the T-matrix of the transistor are obtained from (4.21) and (4.22).

$$B_{CS} = \frac{1 + R_G S(c_{gs} + c_{gd})}{c_{gd} S - g_m} \tag{4.21}$$

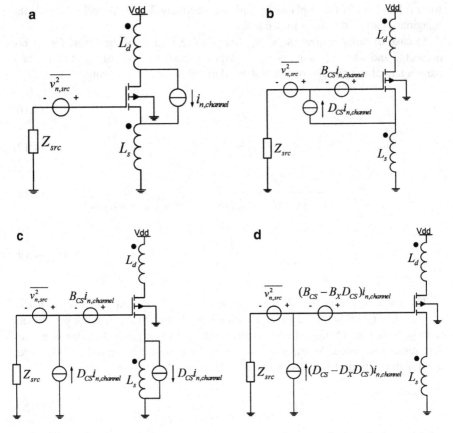

Fig. 4.6 Manipulating the internal noise source of the transistor to refer it to the input, (**a**) the original circuit schematic with the channel noise current source, (**b**) using T-parameters of the transistor to refer the channel noise to the gate of the transistor, (**c**) splitting the floating noise current source between gate and source, (**d**) using T-parameters of the whole circuit to refer the remaining current source to the input

$$D_{CS} = \frac{(c_{gs} + c_{gd})s}{c_{gd}s - g_m} \tag{4.22}$$

where R_G is the gate resistance of the transistor. The B and D elements of the T-matrix of the other two-port are derived from the following:

$$B_x = \frac{1 + c_{gd}s^2(1 - k_c^2)L_D}{(1 - k_c^2)L_D c_{gd}s^2(c_{gs}s + g_m) + (c_{gs} - nk_c c_{gd})s + (1 + nk_c)g_m} \tag{4.23}$$

$$D_x = \frac{(1 - k_c^2)L_D c_{gd}s^2(c_{gs}s + g_m) + (c_{gs} + c_{gd})s}{(1 - k_c^2)L_D c_{gd}s^2(c_{gs}s + g_m) + (c_{gs} - nk_c c_{gd})s + (1 + nk_c)g_m} \tag{4.24}$$

which in case of unilateralization and meeting the conditions of (4.14) are simplified to:

$$B_x = \frac{1}{(1+n)g_m} \tag{4.25}$$

$$D_x = \frac{(c_{gs} + c_{gd})s}{(1+n)g_m} \tag{4.26}$$

The total noise contribution referred to the input of the circuit is then obtained from:

$$\overline{v_{n,i}^2} = 4KTR_G + |(B_{CS} - B_x D_{CS}) + (D_{CS} - D_x D_{CS})Z_{src}|^2 4KT\gamma g_m \tag{4.27}$$

where Z_{src} is the source impedance. Substituting the values of B and D from (4.21)–(4.22) and (4.25)–(4.26) in (4.27), the input-referred noise is derived as a function of circuit parameters:

$$\overline{v_{n,i}^2} = 4KTR_G + \left(1 + \frac{(c_{gs} + c_{gd})^2 \omega^2}{g_m^2(1+n)^2}\right)|1 + Z_{src}(c_{gs} + c_{gd})j\omega|^2 \times \frac{4KT\gamma}{g_m} \tag{4.28}$$

The noise figure is then calculated from the following:

$$NF = 1 + \frac{R_G}{R_{src}} + \frac{\gamma}{g_m R_{src}}\left(1 + \frac{(c_{gs} + c_{gd})^2 \omega^2}{g_m^2(1+n)^2}\right)|1 + Z_{src}(c_{gs} + c_{gd})j\omega|^2 \tag{4.29}$$

where R_{src} is the real part of the source impedance. According to (4.29), the NF can be minimized if the reactive part of the source impedance resonates with the gate-source and gate-drain capacitances (c_{gs} and c_{gd}). It is also deduced from (4.15) that the imaginary part of the input impedance is cancelled out when L_G resonates with c_{gs} and c_{gd}. The imaginary part of the input impedance and the simulated NF versus L_G are shown in Fig. 4.7, showing that the minimum noise figure is achieved when the imaginary part of the input impedance is zero. Figure 4.8 shows the calculated noise figure from (4.29) for some typical component values, showing sharp increase in NF for small values of R_{src} and a relatively flat curve for larger values of R_{src}.

4.1.3 Circuit Design

A differential variant of the discussed LNA topology is designed, as shown in Fig. 4.9. The main design objectives, aside from stability discussed before, are low noise figure and high gain. Both are functions of the transistor biasing and width,

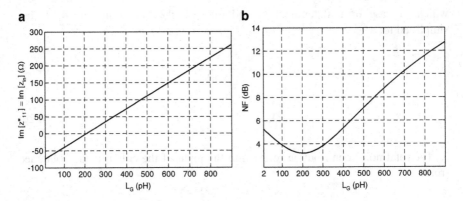

Fig. 4.7 (**a**) Simulated imaginary part of the input impedance of the LNA of Fig. 4.2, (**b**) simulated noise figure of the same LNA versus L_G

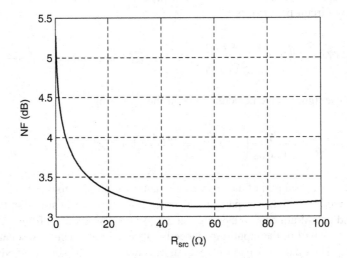

Fig. 4.8 Calculated noise figure of the LNA versus source resistance

passives choices, and source impedance Z_{src}. The single-ended transducer power gain of the LNA after unilateralization is obtained from (4.30).

$$
G_T = \frac{g_m^2 L_D^2 n^2 (1+n) j\omega}{c_{gd}} \times \frac{1}{n^2 + g_m L_D (1+n) j\omega - L_D n (c_{gd} + c_{gs})\omega^2}
$$
$$
\times \frac{1}{n^2 R_L + L_D (n^2 + g_m(1+n)R_L) j\omega - L_D n(c_{gd} + c_{gs})R_L \omega^2}
$$
$$
\times \frac{1}{1 + (c_{gd} + c_{gs}) j\omega (R_G + R_S) - (c_{gd} + c_{gs}) L_G \omega^2} \tag{4.30}
$$

Fig. 4.9 Circuit schematic of the differential voltage–voltage transformer-feedback LNA

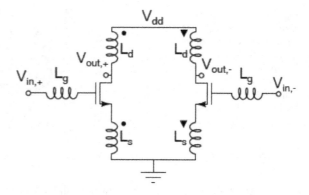

Fig. 4.10 Transformer structure. For clarity the vias connecting the two bottom metals are only shown at the beginning and the end of the metal strips

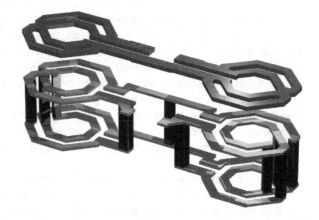

where R_L is the load resistance (not shown in Fig. 4.9). From the last term of (4.30) it is clear that the resonance between L_G and $c_{gs} + c_{gd}$ maximizes the transducer gain, a condition similar to the one required for noise figure optimization and canceling the imaginary part of the input impedance.

The transformer used in the LNA is designed using EM simulations with ADS Momentum. The resulting structure is shown in Fig. 4.10. The top inductor (L_S) consists of two metal lines in parallel to lower the inductance and increase the Q-factor. The bottom inductor (L_D) has two turns, connected through vias which are distributed all along the metal lines. Both inductors are placed exactly on top of each other to achieve the highest possible coupling ($|k_c| \approx 1$). The width of the metal lines is chosen to be 3 μm. This decision constitutes a trade-off between Q-factor and resonance frequency [49]. To satisfy (4.14) for isolation, a turn ratio n of 1.8 is chosen along with a coupling factor k_c of 0.76. The simulated Q-factors of the inductors are higher than 10 at the frequency of interest. Simulated values for L_D and L_S are 13 and 42 pH, respectively. A patterned shield is placed underneath the transformers to reduce substrate coupling.

Two stages of the circuit schematic of Fig. 4.9 are cascaded together to achieve an acceptable gain. The layout of the core of the LNA is shown in Fig. 4.11.

Fig. 4.11 Layout of the LNA
(330 × 170 μm): Only the
top metal layers are shown
to clarify the structure.
Patterned shields are not
shown. The input and
output reference planes
are indicated by *dashed
lines.*

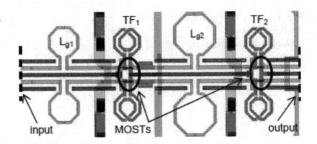

The differential input of the first stage is shown on the left and the differential
output of the second stage is on the right. The two stages are connected to each other
with a DC-blocking capacitor between the output of TF1 and the input of L_{g2}.
All the RF interconnects longer than 10 μm are simulated in ADS Momentum.
The rest of the layout is simulated after RC-extraction. L_{g1} and L_{g2} are approximately 110 pH and 150 pH, respectively. The transistors are indicated in Fig. 4.11
and are situated underneath the metal lines connecting the transformer structures.
Transistor width is 35 and 25 μm in stage 1 and stage 2, respectively. The vertical
lines surrounding the transformers are the DC power lines and biasing of the
LNA. Coplanar waveguides (CPW) with shielding are used to connect different
components to each other. This results in low coupling to the substrate and between
components.

The input and output of the LNA are connected to bondpads using CPWs
(see Fig. 4.12). This results in losses and an impedance shift. The resulting source
and load impedance of the circuit at the input and output indicated in Fig. 4.11 is
approximately $37 + j10\ \Omega$. Open, short, and load structures are added to de-embed
the circuit. A special effort has been put into making the design as symmetrical as
possible to reduce the common mode.

4.1.4 Measurement Results

The LNA, fabricated in CMOS 65 nm technology, is measured using a differential
measurement setup, described in Chap. 3. The DC power consumption is observed
to be equal to the simulated value of 35 mW [16].

The S-parameters are measured using Agilent E8361A power network analyzer.
Calibrations are verified using WinCal XE software. After de-embedding the G_T
calculated for $Z_{src} = 30\ \Omega$ based on the measured s-parameters is 10 dB at 61 GHz
(Fig. 4.13). The measured in-band deviation is ±0.25 dB. The output-to-input
isolation indicated by s_{12} is below −47 dB over the entire measured band of
55–67 GHz and the group delay is close to 20 ps over the entire band of interest.
The differential stability factor (K-factor) remains above 30 in the measured band.

Fig. 4.12 Die photo including the whole LNA with bondpads on the *left* and one de-embedding structure on the *right*. The size of the die is 960 × 980 μm and the size of the LNA is 330 × 170 μm

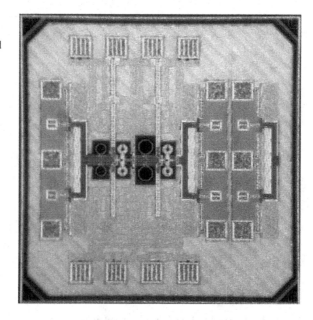

The common mode maximum transducer gain is equal to −2 dB resulting in a CMRR of 12 dB. The s_{12} is below −42 dB, and K-factor remains above 70.

The noise figure is measured in the band 59.5–66 GHz (Fig. 4.13). Z_{src} during this measurement is equal to $37 + j10\ \Omega$, while the input reflection coefficient for the noise source is below −15 dB. The average measured NF in this band is equal to 3.8 dB. Figure 4.14 shows the variations of NF and G_T for different values of source impedance using a source-pull measurement setup. It shows that the NF varies slightly with source impedance such that for a 50 Ω source the NF is only 0.1 dB worse. The NF_{min} of the circuit is found to be 3.7 dB. During this measurement the source impedance for NF_{min} is also verified with the simulated value.

Comparing the measured NF in Fig. 4.14 with the calculated NF in Fig. 4.8, shows a correct prediction of the variation trend of NF with R_{src} by (4.29). Nevertheless the absolute values do not match, because first of all the component values are not the same and secondly (4.19) is derived from a simple model for a single-ended single-stage voltage–voltage transformer-feedback LNA, whereas in Fig. 4.14 we are dealing with a double-stage differential LNA involving all the layout-associated parasitic effects and non-idealities.

The measured IIP3 is equal to 5 dBm at 57.5 GHz and 4 dBm at 60 GHz which is in close agreement with the simulation. The measured 1 dB compression point is −4.6 dBm.

The performance of the existing 60 GHz LNAs is compared with this work in Table 4.1. The LNAs presented in Refs. [41, 50, 51], and Ref. [52] are single ended, and Ref. [53] has a differential output. The work presented here shows the lowest NF and highest bandwidth. The relatively low gain is due to using only two cascaded stages.

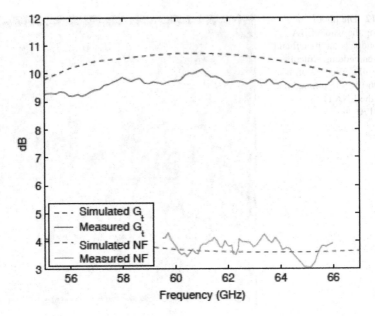

Fig. 4.13 Measured and simulated gain and noise figure

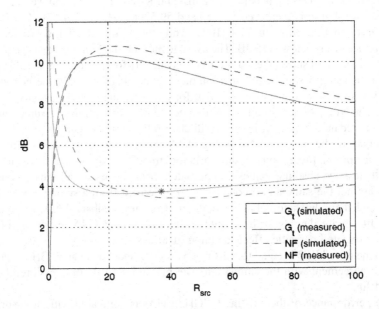

Fig. 4.14 NF and G_T variation as a function of R_{src}, the star in the figure indicates the measurement result without use of the load-pull setup

Table 4.1 Comparison with other works

References	Process (nm)	Topology	G_T (dB)	NF (dB)	3 dB BW (%)	IIP3 (dBm)	V_{DD} (V)	P_{DC} (mW)
[50]	90	3 stage CS	15	4.4	10	N/A	1.3	4
[51]	65 (SOI)	2 stage casc.	12	8	22	N/A	2.2	36
[41]	90	2 stage casc.	14.6	<5.5	25	−6.8	1.5	24
[53]	65	2 stage casc. +1 CS	22.3 (A_v)	6.1	13	N/A	1.2	35
[52] (LNA + mixer)	45	2 stage casc.	26	6	N/A	−12	1.1	23
This work (source 37 + j10)	65	2 stage CS	10	3.8	37	4	1.2	35
This work (source 100 Ω)	65	2 stage CS	8	4.5	–	–	1.2	35

4.2 Mixer

As a critical block of high frequency receivers, mixers pose several challenges to the designers of 60 GHz zero-IF receivers, including DC offset, flicker noise and low-frequency second order intermodulation distortions (IMD2) [54]. Other important general parameters like IP3, conversion gain, noise figure, and power consumption should also be considered in the design flow. The difficulties of 60 GHz design, as compared to low-frequency design, should also be dealt with. The importance of parasitic effects at 60 GHz is such that the designer needs to iteratively shift the focus from schematic to layout and vice versa.

In this work an active double balanced mixer is designed which considers all the above-mentioned issues with emphasis on homodyne challenges. First the process of topology selection is discussed. Afterwards, the theoretical analysis of the selected topology for gain, noise, and nonlinearity is presented. Then the circuit design and the layout of the selected topology are explained. At the end the measurement results are presented.

4.2.1 Topology Selection

The Gilbert cell is one of the most commonly used architectures in mixer design [55]. However, the conventional Gilbert cell does not address zero-IF challenges such as flicker noise and IP2. In fact, contradicting requirements are encountered during the design of a Gilbert cell for homodyne applications. The main dilemma arises from the equality of the biasing current of the transconductor and that of the

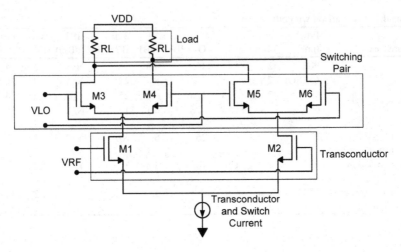

Fig. 4.15 The circuit schematic of a conventional Gilbert cell mixer

switching pair, as shown in Fig. 4.15. Increasing the biasing current which is quite desirable for improving the performance of the transconductor, can be detrimental to the performance of the switching pair in terms of flicker noise [56]. On the other hand the IMD2 generated in the transconductor can easily leak to the output in the presence of mismatched switching pair transistors. To address the former, the current-bleeding topology is proposed to decouple the biasing currents of the transconductor and switching pair, as shown in Fig. 4.15 [57–59]. In this topology the biasing currents of the switching pair and transconductor can be assigned independently. In addition, only a small part of the biasing current of the transconductor flows into the resistive load. Therefore, larger resistors can be used in the load without taking the risk of driving the active transistors into the triode region. Using larger resistors then leads to a higher conversion gain. The additional inductance improves the conversion gain by tuning out the parasitic capacitances in the common-source node.

To address the second problem, the leakage of the IMD2 to the output, many designers opt to isolate the transconductor from the switching pair by an RF-coupling capacitor [60]. In fact, careful inspection of Fig. 4.16 reveals that the transconductor in combination with the inductor forms an independent gain stage which makes no contribution to the task of downconversion. As a result, separating that stage from the switching pair via a capacitor leaves out the switching pair as an independent stage solely responsible for the downconversion, as shown in Fig. 4.17. The omitted part of the circuit with the transconductor can be combined and optimized together with the LNA to provide the required gain. Although the circuit topology of Fig. 4.17 does not include the transconductor, it is still capable of providing power and voltage conversion gain as explained in the next section.

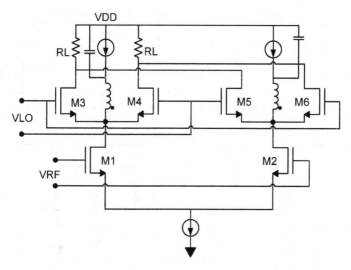

Fig. 4.16 The circuit schematic of a current-bleeding mixer

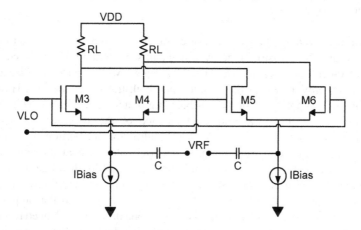

Fig. 4.17 The circuit schematic of an ac-coupled mixer

4.2.2   Gain and Noise Analysis

Several analytical models have been proposed in the literature to clarify the noise and gain mechanisms in the switching pair of the mixers [56, 61, 62]. However, these models are either unable to predict the LO-frequency-dependency of the switching pair noise [56], or are not simple enough for insightful manual calculations in the design phase [61]. In addition, some of them are based on an assumption which loses its validity as the frequency of operation is increased: the noise contribution of the switching pair transistors is assumed to be limited to the moments when both transistors are ON.

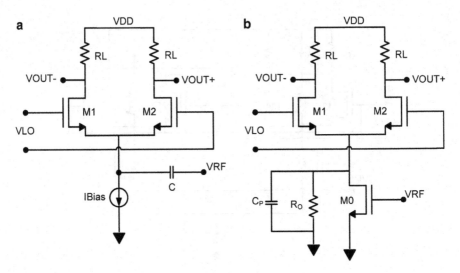

Fig. 4.18 Downconverters in which the current-driven assumption is not valid or applicable: (**a**) ac-coupled mixer and (**b**) Gilber-cell mixer including the common-source capacitance and the transconductor output impedance

In fact in conventional analytical noise models it is implicitly assumed that the source of each ON transistor in the switching pair sees an infinite impedance except when the other transistor is also ON. Consequently, the channel noise of the transistor can only appear at the output when both transistors are ON. However, this assumption is not valid for current-bleeding or ac-coupled mixers (Fig. 4.18a) or for the cases where the output impedance of the previous stage or the transconductor is finite. It even loses its validity for conventional Gilbert-cell mixers as the frequency of operation is increased and the parasitic capacitances of the switch transistors become more influential (Fig. 4.18b). In all these cases, the channel thermal noise of the transistors has always a path to flow from the output load to the ground and therefore each transistor contributes to the output noise whenever it is ON. In this section the analysis is done on single-balanced mixers for simplicity, knowing that the results can be applied to double-balanced mixers with minor modifications.

In some cases, for example at high frequencies, the additional impedances, such as parasitic capacitances, may dominate the impedance seen through the source of each transistor. In these cases the time-varying impedance of the transistors can be neglected as compared to the additional constant impedances, allowing a much simpler time-invariant analysis as presented here. The presented analysis also predicts the dependence of the switching pair noise on the LO frequency.

The circuit of Fig. 4.18a is a generic representation of the switching pair and is used in the rest of this section for gain and noise analysis. Considering each transistor working like a common-gate amplifier when turned on, this circuit can be modeled as in Fig. 4.19 for voltage conversion gain analysis. C is the RF-coupling capacitor also present in Fig. 4.18a and C_S is the accumulated capacitance in the

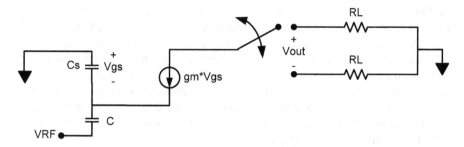

Fig. 4.19 The equivalent circuit used for voltage-conversion gain analysis

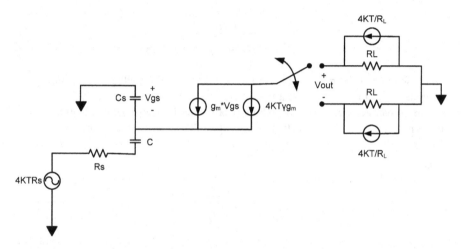

Fig. 4.20 The equivalent circuit used for power-conversion gain and noise analysis (C is the RF-coupling capacitor)

common-source node of M1 and M2 which includes the gate-source capacitance of the transistors and also the source-bulk capacitors. For power-conversion gain calculation we also need the source resistance which is added in Fig. 4.20. Noise sources are also added in Fig. 4.20 for noise analysis. Simplifying assumptions made for performing the calculations are:

1. In each LO semiperiod, the transconductance (g_m) and hence the noise contribution of one transistor is dominant.
2. The g_m of the transistor, which is being turned off, is negligible compared to the capacitive conductance in the common-source node of the switching pair (C_S in this case).
3. The operating points of the transistors remain constant in each LO semiperiod.

The second assumption is more realistic at higher frequencies, but it does not mean that the operation frequency needs to be close to the cutoff frequency, because the g_m of the transistor which is going off is much smaller than that of the ON transistor. The third assumption is less valid for large values of RF-coupling capacitance (C), as it conducts a large time-varying current generated by a sinusoidal LO, modifying the operating point of the transistors significantly. The advantage of using these assumptions as compared to more complicated accurate analyses is the simplicity of the resulting calculations which can facilitate the understanding of gain and noise mechanisms. In fact they provide us with simple and accurate enough relationships for design pre-calculations. To complete the analysis, the effect of sinusoidal LO amplitude, which is initially neglected due to the aforementioned assumptions, is investigated via simulations.

By inspection of the circuit in Fig. 4.19, the voltage conversion gain from V_{RF} to V_{OUT} is calculated as:

$$A_V = \frac{2}{\pi} \frac{g_m C(\omega_{LO} + \omega_{IF}) R_L}{\sqrt{g_m^2 + (C + C_S)^2 (\omega_{LO} + \omega_{IF})^2}} \tag{4.31}$$

which shows that increasing g_m, C, and R_L increases the voltage conversion gain whereas the LO frequency has a negligible effect if C is large enough.

The conversion gain from the available power at the input to the delivered power to the resistances in the load (R_L), using Fig. 4.20, is calculated as follows:

$$G_C = \frac{1}{\pi^2}$$

$$\times \frac{8 R_S R_L g_m^2 C^2 (\omega_{LO} + \omega_{IF})^2}{\left[g_m - C_S C R_S (\omega_{LO} + \omega_{IF})^2 \right]^2 + ((1 + g_m R_S)C + C_S)^2 (\omega_{LO} + \omega_{IF})^2} \tag{4.32}$$

which in case of large C simplifies to:

$$G_C = \frac{1}{\pi^2} \times \frac{8 R_S R_L g_m^2}{C_S^2 R_S^2 (\omega_{LO} + \omega_{IF})^2 + (1 + g_m R_S)^2} \tag{4.33}$$

According to (4.32) and (4.33), the switching pair can provide a power conversion gain which is proportional to the load impedance and the switch transconductance. On the other hand increasing the LO frequency or C_S reduces the power conversion gain, i.e., at higher frequencies it is more difficult to obtain power gain from the switching pair.

The main noise sources that contribute to the output noise of the mixer are: the noise from the input stage or previous stage represented here by the source resistance (R_S), the noise from the switching pair transistors, and the noise of the load resistance (R_L). The noise coming from the LO has the same transfer function to the output as the noise coming from the source resistance, because, as shown in Fig. 4.20, they fall in similar geometrical circuit positions. In the following, the switched common-gate amplifiers model, shown in Fig. 4.20, is used to analyze the contribution of each of these noise sources to the output noise.

The noise of the source resistance from all harmonics of the LO is downconverted to IF at the output. However the noise downconverted by higher harmonics of the LO is smaller than that downconverted by lower harmonics. By turning off all the noise sources in Fig. 4.20, except the noise of the source resistance, the noise going to the output from the source resistance can be calculated:

$$\bar{v}^2_{nout,Rs} = \frac{16KTR_SR_L^2}{\pi^2} \times \sum_{k=0}^{\infty} \frac{1}{(2k+1)^2}$$

$$\times \frac{g_m^2 C^2((2k+1)\omega_{LO} \pm \omega_{IF})^2}{\left[g_m - C_SCR_S((2k+1)\omega_{LO} \pm \omega_{IF})^2\right]^2 + ((1+g_mR_S)C + C_S)^2((2k+1)\omega_{LO} \pm \omega_{IF})^2}$$

$$(4.34)$$

which in case of large C simplifies to:

$$\bar{v}^2_{nout,Rs} = \frac{16KTg_m^2 R_S R_L^2}{\pi^2}$$

$$\times \sum_{k=0}^{\infty} \frac{1}{(2k+1)^2} \times \frac{1}{C_S^2 R_S^2((2k+1)\omega_{LO} \pm \omega_{IF})^2 + (1+g_mR_S)^2} \quad (4.35)$$

and if the higher harmonics are negligible, results in:

$$\bar{v}^2_{nout,Rs} = \frac{16KTg_m^2 R_S R_L^2/\pi^2}{C_S^2 R_S^2(\omega_{LO} + \omega_{IF})^2 + (1+g_mR_S)^2}$$

$$+ \frac{16KTg_m^2 R_S R_L^2/\pi^2}{C_S^2 R_S^2(\omega_{LO} - \omega_{IF})^2 + (1+g_mR_S)^2} \quad (4.36)$$

where K is the Boltzmann constant, T is the temperature, and g_m is the transconductance of the switch transistor in the ON state. According to (4.34), (4.35), and (4.36) the same parameters which increase the conversion gain, also magnify the transfer of noise from the source resistance or the previous stage to the output, which is consistent with our general expectation. Notice that the second term in (4.36) is associated with the image noise.

In conventional analytical models the noise contribution of the switching pair is confined to the simultaneously ON status of the transistors. However, the increasing impact of parasitics (as a result of increasing frequency) and finite output impedance of the previous stage undermine the validity of this assumption. In the proposed analysis each transistor contributes to the output noise in the whole period of its operation in the ON state, irrespective of the other transistor's situation. If the other transistor is also ON, the amount of contribution can change due to the fact that the impedance seen by the source of the transistor is different. Nevertheless, at high frequencies the impedance seen from the source of the transistor which is being turned off is dominated by the common-source node capacitance and hence the effect of its g_m-which is small when the transistor is being turned off- can be neglected. Therefore, in this analysis, we calculate the switching pair noise assuming that in each LO semiperiod one transistor is the dominant noise contributor and that the transistor operation point and the impedance it sees from its source is constant in each LO semiperiod. Turning off all the noise sources in Fig. 4.20, except that of the switching pair, the noise going to the output from the switching pair can be calculated:

$$\bar{v}^2_{nout,sw} = \frac{16KT\gamma g_m R_L^2}{\pi^2} \times \sum_{k=0}^{\infty} \frac{1}{(2k+1)^2}$$

$$\times \frac{C_S^2 C^2 R_S^2 ((2k+1)\omega_{LO} \pm \omega_{IF})^4 + (C+C_S)^2 ((2k+1)\omega_{LO} \pm \omega_{IF})^2}{\left[g_m - C_S C R_S ((2k+1)\omega_{LO} \pm \omega_{IF})^2\right]^2 + ((1+g_m R_S)C + C_S)^2 ((2k+1)\omega_{LO} \pm \omega_{IF})^2}$$

$$(4.37)$$

which in case of large C simplifies to:

$$\bar{v}^2_{nout,sw} = \frac{16KT\gamma g_m R_L^2}{\pi^2}$$

$$\times \sum_{k=0}^{\infty} \frac{1}{(2k+1)^2} \times \frac{C_S^2 R_S^2 ((2k+1)\omega_{LO} \pm \omega_{IF})^2 + 1}{C_S^2 R_S^2 ((2k+1)\omega_{LO} \pm \omega_{IF})^2 + (1+g_m R_S)^2} \quad (4.38)$$

and if the higher order harmonics are negligible, results in:

$$\bar{v}^2_{nout,sw} = \frac{16KT\gamma g_m R_L^2}{\pi^2} \times \frac{C_S^2 R_S^2 (\omega_{LO} + \omega_{IF})^2 + 1}{C_S^2 R_S^2 (\omega_{LO} + \omega_{IF})^2 + (1+g_m R_S)^2} + \frac{16KT\gamma g_m R_L^2}{\pi^2}$$

$$\times \frac{4C_S^2 R_S^2 (\omega_{LO} - \omega_{IF})^2 + 1}{4C_S^2 R_S^2 (\omega_{LO} - \omega_{IF})^2 + (1+g_m R_S)^2} \quad (4.39)$$

According to (4.37), (4.38), and (4.39) the output noise due to the channel thermal noise of the switching pair transistors is proportional to their g_m when they are ON and also to the square of the load resistance. On the other hand the conversion gain is proportional to the square of g_m and R_L. Therefore, increasing g_m is beneficial for reducing the contribution of the switching pair noise to the total noise figure whereas changing R_L does not affect this part of the noise figure.

The noise due to the load resistors at the differential output is given by:

$$\bar{v}^2_{nout,RL} = 8KTR_L \quad (4.40)$$

which means that the output noise power due to R_L is proportional to R_L. However the power conversion gain is proportional to the square of R_L. Therefore, increasing R_L can in fact reduce the contribution of R_L to the total noise figure.

The single-sideband noise figure of the mixer is equal to the ratio of the total output noise to the output noise due to the source resistance noise at the fundamental RF frequency:

$$NF = \frac{\bar{v}_{nout,Rs}^2 + \bar{v}_{nout,sw}^2 + \bar{v}_{nout,RL}^2}{KTG_C \times 2R_L} \tag{4.41}$$

where G_C is the power conversion gain of the mixer. For large values of C, NF is simplified to:

$$NF = \left[C_S^2 R_S^2 (\omega_{LO} + \omega_{IF})^2 + (1 + g_m R_S)^2 \right]$$

$$\times \left(\sum_{k=0}^{\infty} \frac{1}{(2k+1)^2} \times \frac{1}{C_S^2 R_S^2((2k+1)\omega_{LO} \pm \omega_{IF})^2 + (1 + g_m R_S)^2} + \frac{\gamma}{g_m R_S} \right.$$

$$\left. \times \sum_{k=0}^{\infty} \frac{1}{(2k+1)^2} \times \frac{C_S^2 R_S^2((2k+1)\omega_{LO} \pm \omega_{IF})^2 + 1}{C_S^2 R_S^2((2k+1)\omega_{LO} \pm \omega_{IF})^2 + (1 + g_m R_S)^2} + \frac{\pi^2}{2R_L g_m^2 R_S} \right) \tag{4.42}$$

The contribution of the switching pair to the NF can be reduced by increasing the biasing current and g_m, as shown in Fig. 4.21, and the contribution of the load resistance to the NF can be reduced by increasing g_m, as shown in Fig. 4.21, and also by increasing R_L. On the other hand, increasing the frequency or the gate-source capacitance increases the contribution of both the switching pair and R_L to the NF. Figure 4.21 shows the calculated contribution of the switches, R_S, and R_L to the total NF as a function of the biasing current. In calculations, R_L is 300 Ω, the width of each of the switching pair transistors is 32 μm, and the gate length is 65 nm.

The LO amplitude, which has been neglected so far, influences the fraction of LO period in which both switching pair transistors are ON. Ideally the switching takes place instantly which means as soon as one transistor is turned on, the other goes off immediately. However in practice it is not the case and the fraction of LO period in which both transistors are ON can be made smaller by increasing the LO amplitude or using a square-wave LO. Based on our analysis so far, each switching pair transistor contributes to the output noise whenever it is on, regardless of whether the other transistor is ON or off. Therefore, we expect negligible influence by the LO amplitude on the noise going from the transistors to the output. The simulation shown in Fig. 4.22 verifies this conclusion as it shows the small effect of LO amplitude on the simulated output noise. Although the LO amplitude is swept almost rail-to-rail, the output noise variations are less than 1 dB, whereas the gain and noise figure variations are more than 6 dB.

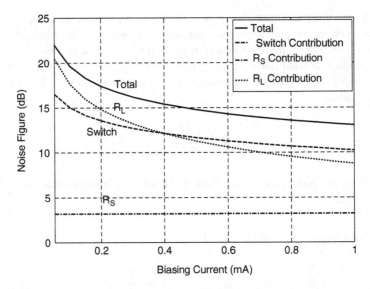

Fig. 4.21 Contribution of each component to the total NF versus biasing current

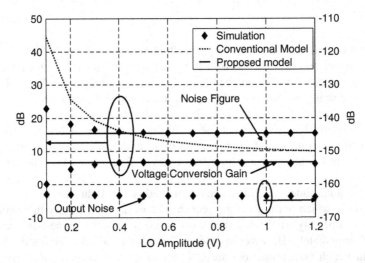

Fig. 4.22 Simulated output noise, noise figure, and voltage conversion gain as a function of LO amplitude; the proposed model fit to the simulation for LO amplitude variations

When both switching pair transistors are ON, they both amplify the input voltage to their outputs, raising the common-mode rather than the differential output. Therefore as the fraction of LO period in which both transistors are ON is prolonged, the conversion gain from the input to the differential output is reduced.

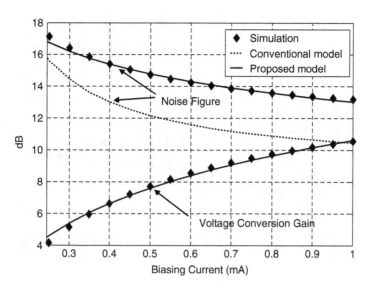

Fig. 4.23 Simulated and calculated noise figure and voltage conversion gain as a function of biasing current

To verify the calculation results by simulations the same parameters used for calculation are used in a single-balanced mixer simulated by SPECTRE-RF periodic steady-state analysis, periodic ac, and periodic noise analyses. The 65 nm TSMC models are used for simulations. The default values are LO frequency of 60 GHz, R_L of 600 Ω, sinusoidal LO amplitude of 600 mV, and biasing current of 400 μA. To introduce the effect of LO amplitude into the calculations, the proposed model is fit by a constant coefficient to the simulated NF and gain versus LO amplitude, so that the simulations and calculations show agreement for moderate and high LO values, as shown in Fig. 4.22. Using the obtained constant coefficient, the simulated and calculated gain and noise figure for different biasing currents are compared in Fig. 4.23. Then in Fig. 4.24 they are compared for different load resistance (R_L) values, and in Fig. 4.25 a comparison is made for different LO frequencies. The simulations and calculations are in good agreement. The IF frequency is chosen at 500 MHz to avoid flicker noise.

The model presented in Ref. [56], does not predict the frequency-dependence of the noise, as shown in Fig. 4.25. According to the conventional model, and as shown in Fig. 4.22, the noise of the switching pair drops significantly as the LO amplitude is increased, because the periods in which both transistors are ON are shortened. However, since the conventional assumption is not realistic at high frequencies, it predicts much higher variations with LO amplitude as compared to reality.

Based on the calculation and simulation results increasing g_m and R_L can improve both gain and noise performance. However, increasing g_m by incrementing

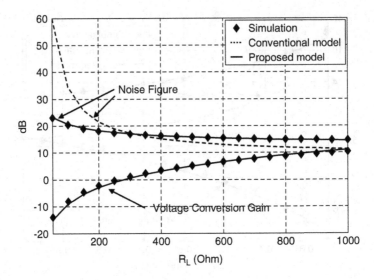

Fig. 4.24 Simulated and calculated noise figure and voltage conversion gain as a function of load resistance

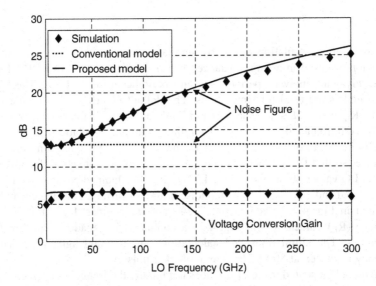

Fig. 4.25 Simulated and calculated noise figure and voltage conversion gain as a function of LO frequency

the biasing current results in two potential problems. First it can drive the transistors into the triode region due to the high voltage drop across the load resistors and second it can increase the flicker noise [56]. Therefore, the optimum biasing current must be determined by considering all these factors.

According to the presented analysis, the contribution of the load resistance to the total NF of a mixer can be reduced by increasing the load resistance. For reducing the contribution of the switching pair to the total NF, wide transistors and high biasing currents are desired. However, combination of high biasing current and high load resistance can drive the switching pair transistors into the undesired triode region. Therefore, there is a limit to improving the noise figure by increasing these two parameters. The fact that each switching pair transistor contributes to the output noise whenever it is ON is verified by simulation results which show negligible variation of the output noise due to LO amplitude variations. However, the LO amplitude affects the conversion gain and hence the noise figure.

4.2.3 Third Order Nonlinearity Analysis

The intermodulation distortion generated in RF mixers can significantly limit the dynamic range of communication systems. The time-varying operation point of the switching pair of active mixers prevents the application of conventional power series and Volterra series method [63, 64]. Therefore, mathematical analyses based on time-varying power series or time-varying Volterra series are proposed to deal with the mixer nonlinearity [65]. However, these mathematical analyses are mostly suitable for use in the software tools and do not yield straightforward closed-form expressions. In this work, the switching pair is described by the combination of a nonlinear time-invariant circuit cascaded with a linear time-varying circuit, allowing the use of conventional power and Volterra series. This approach, previously applied to IMD2 [66], provides closed-form analytical expressions suitable for understanding the underlying mechanisms in the intermodulation process.

The switching pair of an active mixer is split into two parts (shown in Fig. 4.26): a nonlinear time-invariant section which can be analyzed by Volterra series and a linear time-varying section. In fact looking from the output node of the mixer, the switching pair can be regarded as the combination of one transistor, representing M1 and M2 with all their non-idealities and nonlinearities, and an ideal switching pair. We first assume that each transistor in the switching pair is ON for half of the period and the transistors are not ON at the same time. In the course of the analysis it becomes clear that the simultaneous operation of the switching pair transistors in the ON-state has the same impact as the finite impedance in the common-source node has. Therefore, the non-ideal switching behavior can be modeled by an additional parallel resistance in the common-source node of the switching pair. We further assume during the analysis that the operating point of each switching pair transistor is constant during each LO half-period. This assumption is very likely to be violated when a sinusoidal LO is applied in combination with a finite impedance in the common-source node of the switching pair. Nevertheless the presented analysis is capable of capturing the critical mechanisms involved in the

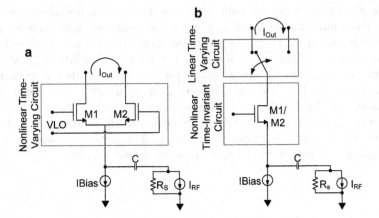

Fig. 4.26 Switching pair of the mixer: (**a**) a nonlinear time-varying circuit, (**b**) transformed into the combination of a nonlinear time-invariant circuit and a linear time-varying circuit. C is the RF-coupling capacitance and R_S represents the source impedance

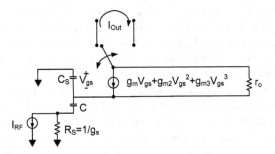

Fig. 4.27 Small-signal model of the switching pair circuit of Fig. 4.26b

intermodulation process. The advantage with respect to more accurate and more complicated models [65] is an easier understanding of the influential mechanisms.

Figure 4.27 shows the small-signal model of the circuit shown in Fig. 4.26b which is used in the rest of the analysis. An inspection of this model reveals that if the capacitance from the common-source node to the ground (represented by C_S) and the source resistance (represented by R_S) are negligible, for example if the operation frequency is low and the previous stage has a very high output impedance, no intermodulation distortion appears in the output, because applying Kirchoff current law (KCL) shows that I_{RF} goes without any change to the ideal switch and gets downconverted. However, the presence of C_S or R_S provides a path for the third-order distortion generated in the voltage-controlled-current-source (VCCS) to flow through the ideal switch. Similarly, one can deduce that the non-ideal switching behavior of the transistors can have the same influence, because if both transistors are ON at the same time, even for a fraction of the whole period, the finite impedance seen from the source of each transistor can provide the path for the distortions generated in the other transistor to flow through the switch to the ground. Therefore, it is possible to model the impact of the non-ideal switching behavior by an additional resistance in parallel with R_S.

Conventional time-invariant Volterra series can be used to analyze the nonlinear time-invariant parts of the circuits of Figs. 4.26 and 4.27. The intermodulation distortion generated in the time-invariant section are downconverted by the ideal linear time-varying switch. First we start with the assumption that the RF-coupling capacitance (C) is big enough to be considered as short-circuit and then we extend the results to the case with small C. Neglecting r_o for simplicity, V_{gs} is expressed as a Volterra series of I_{RF}:

$$V_{gs} = G_1(\omega) \circ I_{RF} + G_2(\omega, \omega) \circ I_{RF}^2 + G_3(\omega, \omega, \omega) \circ I_{RF}^3 + \cdots \tag{4.43}$$

where G_1, G_2, and G_3 are derived in three steps by applying three test inputs of $I_{RF} = e^{j\omega_1 t}$, $I_{RF} = e^{j\omega_1 t} + e^{j\omega_2 t}$, and $I_{RF} = e^{j\omega_1 t} + e^{j\omega_2 t} + e^{j\omega_3 t}$, using Kirchoff current law in the common-source node, and equating the terms with similar frequencies. The intermodulation distortion due to a two-tone interferer $I_{RF} = I_{pk}\sin(\omega_1 t) + I_{pk}\sin(\omega_2 t)$ is derived from (4.44) which is converted to (4.45) after substituting the values of G_1, G_2, and G_3. According to (4.45) there are some terms which can possibly cancel out each other and suppress the whole IMD3. For small values of C the circuit is transformed into a second-order system, as described by (4.46).

$$I_{Out3} = \frac{I_{pk}^3}{2\pi} \times \sin((2\omega_1 - \omega_2 - \omega_{LO})t) \times (3g_m G_3(\omega_1, \omega_1, -\omega_2)$$
$$+ 2g_{m2}G_1(-\omega_2)G_2(\omega_1, \omega_1) + 4g_{m2}G_1(\omega_1)G_2(\omega_1, -\omega_2) + 3g_{m3}G_1^2(\omega_1)G_1(-\omega_2)). \tag{4.44}$$

$$I_{Out3} = \frac{I_{pk}^3}{2\pi} \times \sin((2\omega_1 - \omega_2 - \omega_{LO})t)$$
$$\times \left| \frac{g_s + C_s j(2\omega_1 - \omega_2)}{[g_m + g_s + C_s j\omega_1]^2 [g_m + g_s - C_s j\omega_2][g_m + g_s + C_s j(2\omega_1 - \omega_2)]} \right|$$
$$\times \left| 3g_{m3} - \frac{2g_{m2}^2}{g_m + g_s + 2C_s j\omega_1} - \frac{4g_{m2}^2}{g_m + g_s + 2C_s j(\omega_1 - \omega_2)} \right| \tag{4.45}$$

$$I_{Out3} = \frac{I_{pk}^3 \omega_1^2 \omega_2}{2\pi} \times \sin((2\omega_1 - \omega_2 - \omega_{LO})t)$$

$$\times \left| \frac{1}{\left[-C_s\omega_1^2 + \left(g_m + g_s + \frac{g_sC_s}{C}\right)j\omega_1 + \frac{g_sg_m}{C}\right]^2 \left[-C_s\omega_2^2 - \left(g_m + g_s + \frac{g_sC_s}{C}\right)j\omega_2 + \frac{g_sg_m}{C}\right]} \right.$$

$$\times \frac{g_m}{-C_s(2\omega_1 - \omega_2)^2 + \left(g_m + g_s + \frac{g_sC_s}{C}\right)j(2\omega_1 - \omega_2) + \frac{g_sg_m}{C}}$$

$$\times \left[3g_{m3}\left(\frac{g_m}{C} + j(2\omega_1 - \omega_2)\right) - \frac{2\left(\frac{g_sg_m}{C} + g_{m2}j(2\omega_1 - \omega_2)\right)\left(\frac{g_sg_m}{C} + 2g_{m2}j\omega_1\right)}{-4C_s\omega_1^2 + 2\left(g_m + g_s + \frac{g_sC_s}{C}\right)j\omega_1 + \frac{g_sg_m}{C}} \right.$$

$$\left. - \frac{4\left(\frac{g_sg_m}{C} + g_{m2}j(2\omega_1 - \omega_2)\right)\left(\frac{g_sg_m}{C} + g_{m2}j(\omega_1 - \omega_2)\right)}{-C_s(\omega_1 - \omega_2)^2 + \left(g_m + g_s + \frac{g_sC_s}{C}\right)j(\omega_1 - \omega_2) + \frac{g_sg_m}{C}} \right]$$

$$+ \frac{2g_{m2}\left(\frac{g_sg_m}{C} + 2g_{m2}j\omega_1\right)}{-4C_s\omega_1^2 + 2\left(g_m + g_s + \frac{g_sC_s}{C}\right)j\omega_1 + \frac{g_sg_m}{C}}$$

$$\left. + \frac{4g_{m2}\left(\frac{g_sg_m}{C} + g_{m2}j(\omega_1 - \omega_2)\right)}{-C_s(\omega_1 - \omega_2)^2 + \left(g_m + g_s + \frac{g_sC_s}{C}\right)j(\omega_1 - \omega_2) + \frac{g_sg_m}{C}} - 3g_{m3} \right|$$

$$(4.46)$$

At low frequencies and if the RF-coupling capacitor (C) is removed, the circuit of Fig. 4.27 can be further simplified into a capacitor-less circuit and (4.45) is simplified to:

$$I_{Out3} = \frac{I_{pk}^3}{2\pi} \times \sin((2\omega_1 - \omega_2 - \omega_{LO})t) \times \frac{3g_s}{(g_m + g_s)^4} \times \left(g_{m3} - \frac{2g_{m2}^2}{g_m + g_s}\right) \quad (4.47)$$

If the output resistance of the transistors is also taken into account, assuming the output nodes at the ground potential to avoid ambiguity, (4.47) is converted to (4.48):

$$I_{Out3} = \frac{I_{pk}^3}{2\pi} \times \sin((2\omega_1 - \omega_2 - \omega_{LO})t) \times \frac{3g_s}{\left(g_m + g_s + \frac{1}{r_o}\right)^4}$$

$$\times \left(g_{m3} - \frac{2g_{m2}^2}{g_m + g_s + \frac{1}{r_o}}\right) \quad (4.48)$$

According to (4.45) and (4.48), the third-order intermodulation terms at the output current are generated not only directly by the third-order transconductance (g_{m3}) of the transistors, but also by the second and first-order transconductances transferring the nonlinearity distortions in V_{gs} to the output. This produces two terms in (4.45) and three terms in (4.48) which can potentially cancel out each other. In fact at low biasing currents the values of g_m, g_{m2}, and g_{m3} are positive, thus for some values of biasing current or source resistance, third order intermodulation in (4.45) and (4.48) can be made zero. However, at high biasing currents g_{m3} becomes negative and those terms do not cancel out each other anymore.

To verify the theory with simulations, the circuit of Fig. 4.26a is simulated by SPECTRE-RF periodic-steady-state and periodic-ac analysis at both low and high frequency regimes. The transistor width and transistor length are 32 µm and 65 nm, respectively. First, a 60 kHz square-wave LO signal with an amplitude of 0.6 V and two RF tones at 61 and 61.1 kHz are used. Figures 4.28 and 4.29 show the simulated and theoretical IMD3 current at the output as a function of R_S and I_{Bias}, respectively. It can be seen that the presented analysis captures the trend of variation with an acceptable accuracy. Particularly the values of R_S for which the two terms in (4.48) cancel out each other are predicted with a good agreement by both simulation and theory (occurring between 10 and 100 Ω). Figure 4.28 clearly shows that increasing R_S beyond a certain point reduces the IMD3, because if the common-source node is open for nonlinearity distortions they just form a current loop in r_o and the VCCS, failing to appear in the output. On the other hand for values of R_S smaller than 1 kΩ, IMD3 has an increasing trend with respect to R_S. This is because for small R_S most of the I_{RF} flows into R_S resulting in a small portion of the current absorbed by the transistors.

Also at high frequencies and according to (4.45) there are some terms which can possibly cancel out each other and suppress the whole IMD3. This can be seen in Fig. 4.30 which compares the simulated and theoretical IMD3 versus R_S for I_{Bias} equal to 50, 100, 150, 200, and 250 µA. The simulation is done with the same setup for low frequency, except that the LO is a 60 GHz sinusoidal with 0.6 V amplitude and the RF tones are at 61 and 61.1 GHz with an RF-coupling capacitance (C) of 1 µF. Both theory and simulation show that for some specific values of R_S the IMD3 converges to zero. In simulation this happens between 10 and 100 Ω and in theory this happens between 100 Ω and 1 kΩ. This discrepancy between simulation and theory is attributed to the fact that the LO signal generates a current in the RF-coupling capacitance (C) which modifies the operating point of the transistors significantly and contradicts a basic simplifying assumption of the analysis about constant operating point of the transistors. For small values of C, as suggested by (4.46) and shown in Fig. 4.31, the zeroes in the IMD3-R_S curve disappear.

At such a high frequency and with a sinusoidal LO the transistors are ON together for a significant fraction of each LO period. This effect is taken into account by introducing a constant additional resistance in parallel with R_S into the calculations.

Comparing Fig. 4.28 with Figs. 4.30 and 4.31 shows that at high frequencies the IMD3 is not reduced by increasing R_S. This is because increasing R_S beyond a

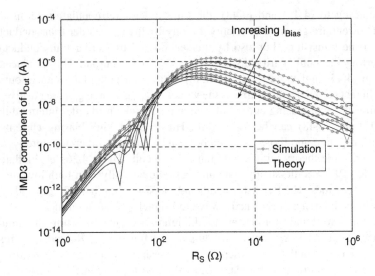

Fig. 4.28 Simulated and theoretical IMD3 of the output current versus R_S, for I_{Bias} equal to 250, 300, 400, 500 and 600 μA at low frequency

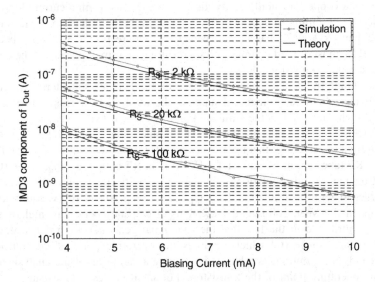

Fig. 4.29 Simulated and theoretical IMD3 of the output current versus I_{Bias} at low frequency

certain point makes it negligible compared to the parasitics of the common-source node (represented by C_S) and thereafter the IMD3 remains unchanged by varying R_S, whereas at low frequencies R_S is the only determining factor of the common-source node impedance.

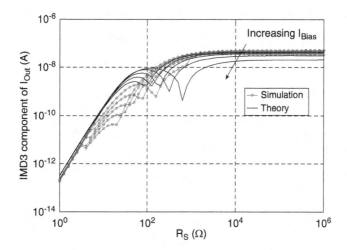

Fig. 4.30 Simulated and theoretical IMD3 of the output current versus R_S for a 60 GHz LO and $C = 1\ \mu F$

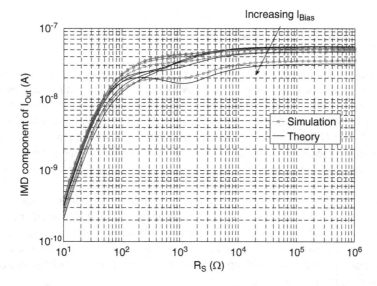

Fig. 4.31 Simulated and theoretical IMD3 of the output current versus R_S for a 60 GHz LO and $C = 1$ pF

Figure 4.32 shows the IMD3 versus biasing current for R_S equal to 10 and 100 kΩ. The variation of IMD3 with respect to biasing current at high frequency is much smaller than the variation at low frequency, shown in Fig. 4.29. This is mostly caused by the fact that by increasing the frequency the bias-independent parasitic capacitances become more influential than the bias-dependent transconductances of the transistors.

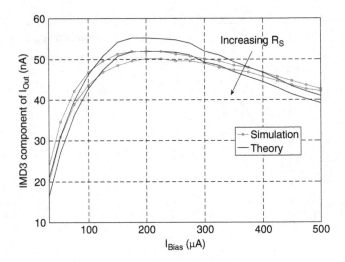

Fig. 4.32 Simulated and theoretical IMD3 of the output current versus I_{Bias} for a 60 GHz LO and $C = 1$ pF

4.2.4   Second Order Nonlinearity Analysis

The second-order intermodulation distortion (IMD2) is one of the major problems in zero-IF receivers. If not suppressed by proper design or some run-time cancellation, it can deteriorate the sensitivity and BER. One possible approach to this problem is to filter out the interferers with highly selective RF filters before they can reach the LNA and mixer. However, bulky RF filters are not practical solutions for integrated receiver front-ends. Therefore, minimization of second order intermodulation distortion remains a challenge in zero-IF receiver design, creating enough motivation for their study and analysis.

The downconversion mixer is normally the main contributor to the IMD2. The low-frequency second-order distortions generated in the LNA or other preceding stages can easily be filtered by RF coupling or band-pass filtering. The mixer of Fig. 4.18a is redrawn in Fig. 4.33 including load capacitors (C_L) which represent the input capacitance of the following stage as well as the parasitics of the switching transistors at the output node.

The differential output IMD2 voltage ($V_{imd2,out}$), comes from two sources: the common-mode output IMD2 current combined with output load mismatch and differential-mode output IMD2 current, as defined in (4.49) and (4.50).

$$I_{imd2Diff} = I_{imd2,1} - I_{imd2,2} \tag{4.49}$$

$$I_{imd2CM} = \frac{I_{imd2,1} + I_{imd2,2}}{2} \tag{4.50}$$

Fig. 4.33 The ac-coupled
mixer used in the IMD2
analysis

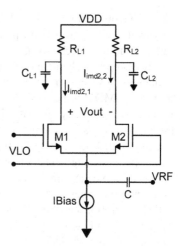

where $I_{imd2,1}$ and $I_{imd2,2}$ are as shown in Fig. 4.32. The differential output IMD2
voltage is described as a function of these currents in (4.51):

$$V_{imd2out} = I_{imd2,1}Z_{L,1} - I_{imd2,2}Z_{L,2} \tag{4.51}$$

where $Z_{L,1}$ and $Z_{L,2}$ are the impedances seen from the Vout$^+$ and Vout$^-$ nodes
respectively:

$$Z_{Li} = \frac{R_{outi}}{1 + R_{outi}C_{Li}j\omega} \quad (i = 1, 2) \tag{4.52}$$

where R_{out} is the resistance seen from the output node.

Defining a nominal value for the output impedance as in (4.53), the differential
output IMD2 voltage can be rewritten as a function of common-mode and differential-mode output IMD currents as depicted in (4.54).

$$Z_{L,1} = Z_L(1 + \delta Z_L)$$
$$Z_{L,2} = Z_L(1 - \delta Z_L) \tag{4.53}$$

$$V_{imd2out} = 2I_{imd2CM}Z_L\delta Z_L + I_{imd2Diff}Z_L \tag{4.54}$$

I_{imd2CM} is a function of the input-stage and switching stage even-order
nonlinearities and is present at the output current even if there is no mismatch in
the circuit. However, it can be vanished in the differential output voltage by a
perfect matching between Z_{L1} and Z_{L2}.

Three mechanisms are responsible for generation of $I_{imd2DIF}$: self-mixing, previous stage nonlinearity combined with switching pair mismatches, and switching
pair nonlinearity combined with its mismatches [66].

Self-mixing is a result of the leakage of the RF signal to LO and vice versa [67].
For instance the RF signal leaked to the LO port can downconvert itself resulting in
a second-order distortion term appearing in the IF output. The leakage is a function

of the layout parameters and is zero in an ideally matched fully balanced downconverter. However, in practice any kind of mismatch in the LO-RF paths can activate this mechanism.

The second mechanism is a consequence of the leakage of the IMD2 generated in the previous stage to the IF output. In an ideally matched switching pair the low-frequency IMD2 of the preceding stage is upconverted and hence falls outside the IF band. However any mismatch in the switching pair transistor allows the IMD2 of the preceding stage to leaks to the output without frequency conversion. In the circuit of Fig. 4.33, proper selection of the RF-coupling capacitance (C) can provide a high-pass filtering which suppresses the IMD2 of the preceding stage.

The third mechanism is associated with the switching pair and appears as a result of its nonlinearities and mismatches. To describe this particular mechanism of IMD2 generation in a typical mixer shown in Fig. 4.34a, the model of Fig. 4.34b can be used. In Fig. 4.34a the mismatch between the two switching pair transistors, M1 and M2, is modeled by an offset voltage source in series with the gate of M1. In an ideal case of square-wave switching, the switching pair can be looked at as a single transistor with a voltage source in series with the gate and toggling between 0 and V_{off} [66]. Then this transistor is connected to the differential load with an ideal switch which alternates between the two output nodes with the same frequency as the LO signal. Taking higher order transconductances of M1 and M2 into account, one can show that if a two-tone (with components at ω_1 and ω_2) signal is applied to the RF input, drain current spectral components are produced at $k\omega_{LO} \pm (\omega_1-\omega_2)$ which are then downconverted by the ideal switch to the baseband, contributing to the output IMD2. The resulting IMD2 is described by [66]:

$$
\begin{aligned}
I_{imd2out} = {} & \frac{I_{RF}^2 \times 2V_{off}(\omega_{LO}\tau)^2}{\pi^2\left[1 + (\omega_1\tau)^2\right]} \\
& \times \sum_{k\in\mathbf{N}} \mathrm{Re}\left\{ \frac{1}{[1 + j(2k + 1)\omega_{LO}\tau]^2} \right. \\
& \left. \times \left[\frac{4g_{m2}^2}{g_m^3}\left(\frac{3 + j4(2k + 1)\omega_{LO}\tau - 4k(k + 1)(\omega_{LO}\tau)^2}{1 + j2(2k + 1)\omega_{LO}\tau - 4k(k + 1)(\omega_{LO}\tau)^2} \right) + \frac{6g_{m3}}{g_m^2} \right] \right\} \\
& + \frac{I_{RF}^2 \times 16V_{off}(\omega_{LO}\tau)^2}{\pi^2\left[1 + (\omega_1\tau)^2\right]} \times \sum_{k\in\mathbf{N}} \mathrm{Re}\left\{ \frac{k^2}{(4k^2 - 1)}(1 + j2k\omega_{LO}\tau)^2 \right. \\
& \left. \times \left[\frac{4g_{m2}^2}{g_m^3}\left(\frac{3 + j8k\omega_{LO}\tau + (1 - 4k^2)(\omega_{LO}\tau)^2}{1 + j4k\omega_{LO}\tau + (1 - 4k^2)(\omega_{LO}\tau)^2} \right) - \frac{6g_{m3}}{g_m^2} \right] \right\}
\end{aligned}
$$

$$(4.55)$$

where I_{RF} is the current flowing into the switching pair as a result of the RF interferer, ω_1 is the RF interferer frequency, g_m, g_{m2} and g_{m3} are the first, second, and third order transconductances of the switching pair transistors respectively, τ is

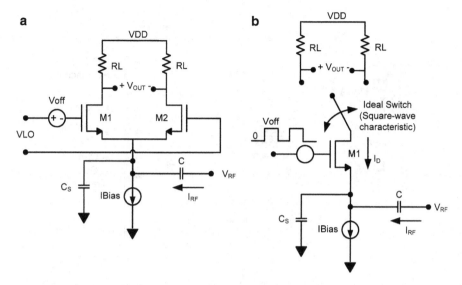

Fig. 4.34 The switching pair with mismatch: (**a**) the mismatch modeled by a voltage offset, (**b**) the switching pair modeled by a single-transistor and an ideal switch

the time-constant of the common-source node of the switching pair, and V_{off} is the equivalent voltage offset representing the mismatch between the transistors [68]:

$$V_{off} = \Delta V_T + \frac{\partial f}{\partial \beta} \Delta \beta + \frac{\partial f}{\partial \theta} \Delta \theta$$

$$f = \frac{\theta}{\beta} I_{DS} \left(1 + \sqrt{1 + \frac{2\beta}{\theta^2 I_{DS}}} \right) + V_T \qquad (4.56)$$

where f is an auxiliary variable with voltage dimension, I_{DS} is the biasing current of the transistor, $\beta = \mu_n C_{ox} W/L$, θ is a fitting factor taking into account the short channel effects, and V_T is the threshold voltage. It is implied by (4.55) that for small values of $\tau \omega_{LO}$, the generated IMD2 is small. Consequently, at low frequencies, for small parasitic capacitance, or large biasing currents, the IMD2 generated in the switching stage is negligible.

4.2.5 Circuit Design and Layout

The double-balance mixer of Fig. 4.17 is designed in CMOS 65 nm technology. The sizing of each switching pair transistor is 10 μm/65 nm, composed of five fingers. The current sources are implemented with NMOS transistors with 4 μm/ 65 nm sizing. V_{DD} is 1.2 V. The load resistances are 2.5 kΩ each and are

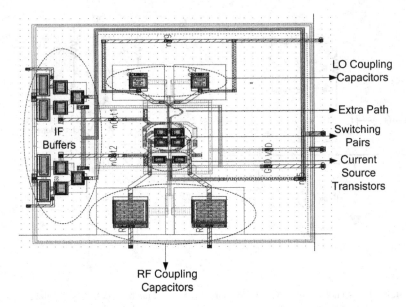

LO Coupling
Capacitors

Extra Path

Switching
Pairs

Current
Source
Transistors

IF
Buffers

RF Coupling
Capacitors

Fig. 4.35 The layout of the core of the mixer including the buffers and RF-coupling capacitors

implemented with poly resistors. The RF-coupling capacitors are implemented with metal-insulator-metal (MIM) capacitors and are 300fF each. Two MIM capacitors are also used at the LO ports to decouple the DC biasing and LO signal. In addition, two source-follower buffers are used at the IF output. Two center-tapped 407 pH spiral inductors are used for RF input matching.

Special attention is paid to the matching of the paths and connections during the layout to minimize IMD2. Figure 4.35 shows the layout of the mixer, zooming in on the core and the output IF buffer. The extra path in one of the LO lines is used to match the length of the two LO connections.

The whole layout including the bondpads and matching network inductors is shown in Fig. 4.36. After distributing the ground lines with Metal layer 1 strips and increasing the number of decoupling capacitances, the layout looks like what is shown in Fig. 4.37. The die photo is shown in Fig. 4.38.

4.2.6 Measurement Results

The circuit is measured on-wafer with the measurement equipments described in Chap. 3. A 60 GHz LO signal with 0 dBm amplitude is used in all the measurements. Figure 4.39 shows the measured high-frequency noise of the mixer versus IF frequency. Figure 4.40 shows the measured power conversion gain. Comparison with simulations shows about 2 dB discrepancy between the results which can be attributed to the post-layout modeling inaccuracies during the simulation.

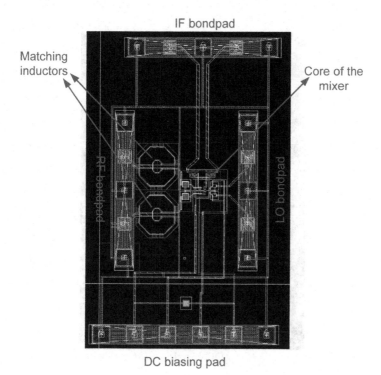

Fig. 4.36 The layout of the mixer including the bondpads and matching network

Figure 4.41 shows the second-order and third-order intermodulation distortion resulting from a two-tone signal. An extrapolation of the results of Fig. 4.41 yields an IP2 and IP3 of 16.6 and −6 dBm, respectively.

Table 4.2 shows the measurement results of the mixer compared with another stand-alone mixer in the literature. In Ref. [69], a conventional Gilbert cell is implemented including a transconductor which is expected to boost the conversion gain. However, the reported voltage conversion gain is similar to this work that uses no transconductor. It is worth mentioning that the IF buffers used in this work are common-source amplifiers with unity-gain feedback and thus do not contribute to the measured voltage conversion gain. The IP3 and 3 dB bandwidth of this work are also superior to the other work by 2 dB and 300 MHz, respectively.

4.3   Quadrature VCO

Generating a quadrature-phase LO signal is essential for direct downconversion of the RF signal in zero-IF receivers. Utilizing a sliding-IF architecture by double conversion of the RF signal with a local oscillator operating at lower frequencies is

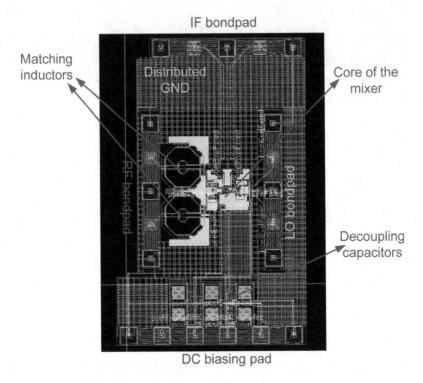

IF bondpad

Matching
inductors

Distributed
GND

Core of the
mixer

Decoupling
capacitors

DC biasing pad

Fig. 4.37 The layout of the mixer including the bondpads, matching network, the distributed ground and the decoupling capacitors

another alternative, but requires more components and filters [70]. In this section, different techniques of quadrature 60 GHz signal generation are investigated and an actively coupled LC quadrature VCO is selected for this purpose [71]. Theoretical analysis of the tuning range and phase noise of the selected topology is explained. Then circuit design, layout, and measurement results are presented.

4.3.1 Quadrature Signal Generation Techniques

Figure 4.42 shows three different ways of generating a quadrature signal at 60 GHz. As shown in Fig. 4.42a a 120 GHz VCO can be used in combination with a quadrature divider-by-two to provide 60 GHz quadrature signals at the output [72]. However, implementing such high-frequency circuits in CMOS, operating at double the required frequency, with the required phase noise and tuning range is not an easy task. The second method shown in Fig. 4.42b simply uses a normal single-phase 60 GHz VCO but shifts the complexities to the following stage by calling for passive phase-shifters which usually suffer from high insertion loss at mm-wave range. The method chosen in this work is designing a quadrature 60 GHz VCO, as shown in Fig. 4.42c.

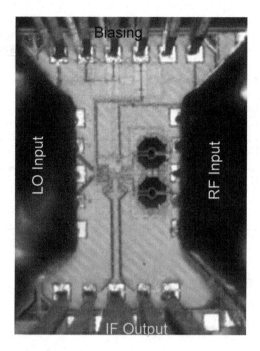

Fig. 4.38 Die photo of the mixer

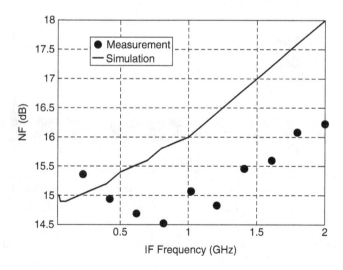

Fig. 4.39 The measured NF versus IF frequency

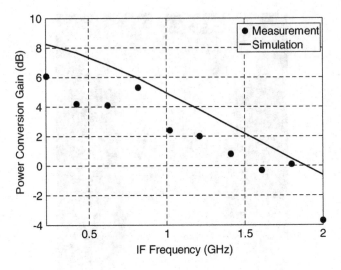

Fig. 4.40 The measured conversion gain versus IF frequency

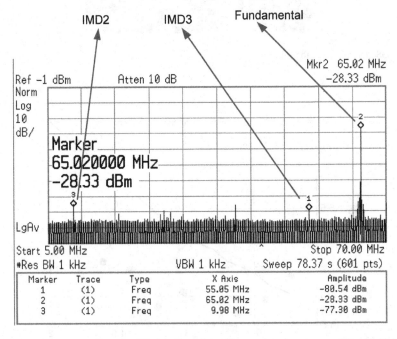

Fig. 4.41 The measured intermodulation distortions after applying a two-tone RF with 10 MHz spacing (only one tone is shown)

Such VCOs are usually designed by establishing some sort of coupling between two cross-coupled pair VCOs, as shown in Fig. 4.43. The coupling can be implemented in different ways such as transformer coupling or active coupling.

Table 4.2 Measurement results of the mixer compared with another work

	[69]	This work
Power conversion gain @ 500 MHz (dB)	–	4
Voltage conversion gain @ 500 MHz (dB)	2	2
NF (SSB) @ 500 MHz (dB)	–	15
DC power (mW)	–	6 (including buffers)
Input impedance (Ohm)	–	40–40j
IIP2 (dBm)	–	16.6
IIP3 (dBm)	–8	–6
IF 3-dB bandwidth (GHz)	1	1.3
Technology	130 nm CMOS	65 nm CMOS

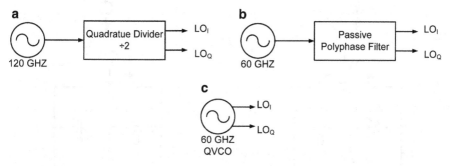

Fig. 4.42 Three different ways of generating a quadrature 60 GHz signal: (**a**) a single phase 120 GHz VCO cascaded with a quadrature divider-by-two, (**b**) a single-phase 60 GHz VCO cascaded with a passive polyphase filter, and (**c**) a quadrature 60 GHz VCO

In transformer coupling, as shown in Fig. 4.43b, the coupling is implemented using inductive coupling between the tank inductor and an inductor used in the source of the cross-coupled transistors [73]. This type of VCO has been designed and simulated at 60 GHz [74, 75]. In this work, the VCO shown in Fig. 4.43c is designed and fabricated. In this topology, the coupling between the cross-coupled pairs is established using additional transistors M5–M8. The cross-coupled pairs M1–M2 and M3–M4 generate the required negative conductance and the coupling transistors M5–M8 inject the output signals of one cross-coupled pair to the input of the other to produce anti-phase coupling required for quadrature generation [71].

4.3.2 Tuning Range

The oscillation frequency of an LC oscillator is given by:

$$\omega = \frac{1}{\sqrt{LC}} \tag{4.57}$$

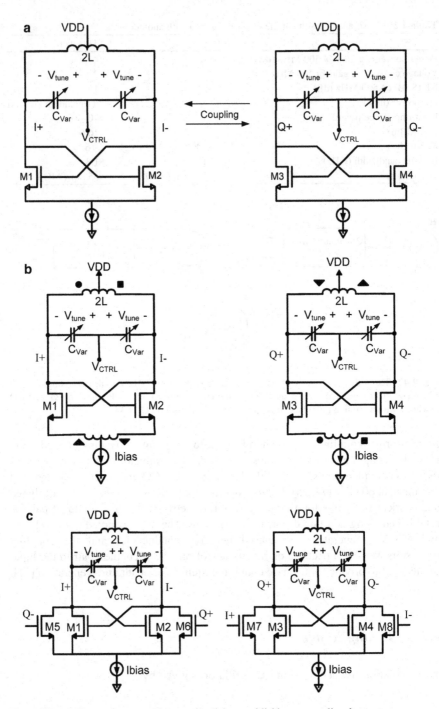

Fig. 4.43 (a) Generating a quadrature signal by establishing a coupling between two cross-coupled pair VCOs with LC tanks, (b) transformer coupling, (c) transistor coupling

where L is the tank inductance and C is the tank capacitance which includes all the parasitic capacitances of the transistors as well the varactors used for frequency tuning.

In fact the varactors are responsible for changing the capacitance in (4.57) and thereby modifying the oscillation frequency. The tuning range of a VCO is a measure of the frequency range covered by the VCO and is defined by:

$$TR(\%) = \frac{\omega_{max} - \omega_{min}}{\omega_{center}} \times 100 \qquad (4.58)$$

where ω_{max}, ω_{min}, and ω_{center} are the maximum, minimum, and center frequency of oscillation, respectively. The maximum and minimum frequencies correspond to the lowest and highest values of the varactor capacitance, respectively. The center frequency, ω_{center}, is equidistant from both ω_{max} and ω_{min}. Therefore, the tuning range can be described also as:

$$TR(\%) = \frac{\omega_{max} - \omega_{center}}{\omega_{center}} \times 200 \qquad (4.59)$$

or

$$TR(\%) = \frac{\frac{1}{\sqrt{LC_{min}}} - \omega_{center}}{\omega_{center}} \times 200 \qquad (4.60)$$

where C_{min} is the minimum capacitance attainable by the tank capacitance as a result of setting the varactor capacitance to its minimum value. The maximum capacitance of the tank, corresponding to the minimum frequency, must then satisfy the following:

$$\frac{1}{\sqrt{LC_{max}}} = 2\omega_{center} - \frac{1}{\sqrt{LC_{min}}} \qquad (4.61)$$

Based on (4.60) and (4.61), increasing the tuning range requires reducing C_{min} or L and increasing C_{max} accordingly. Although the tuning range can be increased to a maximum of 200% based on these equations, in practice there are two factors which limit the obtainable tuning range. First of all, C_{min} cannot be reduced arbitrarily as its minimum value is limited by the parasitic capacitances and the fixed part of the varactor capacitance. Secondly the alternative solution which demands reducing L is problematic at high frequencies. Because at high frequencies the quality factor (Q) of the tank is dominated by the low-Q varactors. As the size of the inductor is reduced and C_{max} is increased accordingly, the loss of the tank is raised and the necessary conditions of oscillation are put at risk. To further explain this point, we have to consider that the cross-coupled pair, i.e. M1–M2 or M3–M4 in Fig. 4.43, gives rise to a negative resistance which is essential for oscillation and is given by:

$$R_{eq} = -\frac{2}{g_m} \qquad (4.62)$$

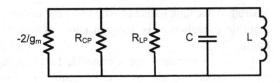

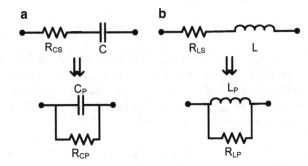

where g_m is the transconductance of each cross-coupled pair transistor. In fact, the combination of a cross-coupled pair and an LC tank, shown in the circuit of Fig. 4.43a, can be modeled by the equivalent circuit of Fig. 4.44. The necessary condition for oscillation is that the positive resistances arising from the losses of the tank are completely nullified by the negative resistance of the cross-coupled pair:

$$\frac{g_m}{2} > \frac{1}{R_{CP}} + \frac{1}{R_{LP}} \tag{4.63}$$

where R_{LP} and R_{LP} are the equivalent parallel resistances associated to the losses of the tank capacitance and inductance, respectively. To calculate R_{CP} and R_{LP}, one has to perform the series-to-parallel transformations of Fig. 4.45 which yields:

$$R_{CP} = \frac{1 + Q_C^2}{Q_C C \omega} \tag{4.64}$$

$$R_{LP} = \frac{L\omega\left(1 + Q_L^2\right)}{Q_L} \tag{4.65}$$

where Q_C and Q_L are the quality factors of the capacitance and inductance, respectively and are given by:

$$Q_C = \frac{1}{R_{CS} C \omega} \tag{4.66}$$

$$Q_L = \frac{L\omega}{R_{LS}} \tag{4.67}$$

Replacing the values of R_{CP} and R_{LP} from (4.64) and (4.65) into (4.63) yields the minimum inductance required for satisfying the oscillation condition:

$$L\omega > \frac{2}{g_m} \left[\frac{Q_C}{1 + Q_C^2} + \frac{Q_L}{1 + Q_L^2} \right] \qquad (4.68)$$

which must be satisfied at all frequencies in the tuning range. Therefore there is a lower limit for the tank inductance which is determined by the quality factor of the capacitor and inductor. Since at high frequencies, the quality factor of a varactor is much lower than that of an inductor, the lower limit of the inductance is mainly determined by the quality factor of the varactors.

Therefore, obtaining the required tuning range from a high-frequency VCO is hindered by both the fixed capacitance of the tank and the low quality factor of the varactors. The former arises from the parasitic capacitances and the fixed part of the varactor capacitance and the latter is a result of the losses of the varactor which are accentuated at high frequencies.

On the other hand, according to (4.68), one can lower the minimum allowable L by increasing g_m. This can be intuitively explained by the fact that increasing g_m magnifies the negative conductance of the cross-coupled pair. That in turn compensates the additional losses of the tank due to additional varactors used for extending the tuning range, because, in fact, reducing the value of L provides more room for additional varactors and in this way extends the tuning range. Increasing g_m can be done in two ways: making the cross-coupled transistors wider or raising the biasing current and power consumption. The former is not desirable as it increases the parasitic capacitances and hence the value of C_{min}. Therefore, the second approach can be used as a way of trading power consumption for tuning range.

4.3.3 Phase Noise

The output voltage of a sinusoidal LC oscillator can be expressed in time domain by:

$$v(t) = A \cos[\omega_0 t + \phi(t)] \qquad (4.69)$$

where A is the amplitude, ω_0 is the oscillation frequency, t is the time. $\Phi(t)$ represents the phase and is commonly treated as a zero-mean stationary random process to include deviations of the phase from the ideal. These random deviations of the phase are caused by the noise of the passive and active components of the VCO. Even the amplitude (A) is susceptible to random deviations as a result of different types of noise which are present in the VCO circuit. As a consequence of the random fluctuations of A and $\Phi(t)$ the output spectrum of the oscillator contains two sidebands of noise power around the oscillation frequency. These

Fig. 4.46 Typical spectrum
of the phase noise of an
oscillator predicted by
Leeson-Cutler model

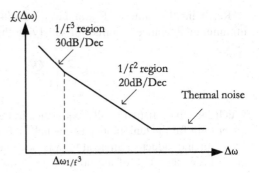

undesired sidebands are best quantified by a parameter called phase noise which is
defined as the total single-sideband output noise normalized to the power of the
sinusoidal output of the oscillator.

The phase noise of the oscillators has been widely studied in the past 50 years.
One of the very first and insightful models of the phase noise, known as Leeson-
Cutler model, was proposed in 1965 and 1966 and expresses the phase noise as
[76–78]:

$$\pounds(\Delta\omega) = 10\log\left[\frac{2FKT}{P_S}\left(1 + \left(\frac{\omega_0}{2Q\Delta\omega}\right)^2\right)\left(1 + \frac{\Delta\omega_{1/f^3}}{|\Delta\omega|}\right)\right] \qquad (4.70)$$

where F is an empirical parameter, called effective noise figure by Leeson, K is the
Boltzmann's constant, T is the absolute temperature, P_S is the average power
dissipated in the resistive part of the tank, ω_0 is the oscillation frequency, Q is the
effective quality factor of the tank, $\Delta\omega$ is the offset from ω_0, and $\Delta\omega_{1/f^3}$ is the
frequency of the corner between the $1/f^3$ and $1/f^2$ regions, as shown in phase noise
spectrum of Fig. 4.46. According to Fig. 4.46, there are three main regions in the
phase noise spectrum. Close to the carrier there is a region which is called $1/f^3$ and
the phase noise drops with a rate of 30 dB/decade. The biggest region in the middle,
called $1/f^2$ region, exhibits a 20 dB/decade falling rate. Eventually, at large frequency offsets, the phase noise spectrum forms a flat region. According to this
simple model, raising the quality factor of the tank, which is directly determined by
the losses of the tank, can improve the phase noise performance. As mentioned
before, the quality factor of the varactors is inferior to that of the inductors at high
frequencies. Therefore, the varactors are the dominant factor in determining the
phase noise resulting from the tank. However, the Leeson-Cutler model remains
silent about the contribution of the active parts to the phase noise. In fact the effect
of the active parts is assumed to be specified by F which strongly depends on the
oscillator topology. Finding the value of F is not addressed by Leeson-Cutler
model. The other simplifying assumption made by this model is considering the
oscillator as a linear time-invariant system, whereas in practice the time-varying
nature of the oscillator can significantly influence its phase noise behavior.

In the past 20 years the design community has significantly progressed beyond the Leeson-Cutler classic linear model and has developed analysis methods that take the time-varying and large-signal nature of the oscillators into account. For example in Ref. [79] an expression is derived for F in cross-coupled pair LC-tank VCOs, incorporating the nonlinear nature of the circuit. According to their analysis, the flicker noise in the tail current source at frequency $\Delta\omega$, upconverts to $\omega_0 \pm \Delta\omega$, and enters the resonator as AM noise (as opposed to PM noise). In the presence of a high gain varactor this AM noise is converted to FM and appears as phase noise. Therefore, it is important to pay careful attention to the sizing of the tail current source transistor to keep its flicker noise lower than required. Another work presented by Hajimiri and Lee puts higher emphasis on the time-variant nature of VCOs and greatly facilitates understanding the process of AM-to-PM noise conversion [80, 81]. Central to their work is defining a function called impulse sensitivity function (ISF) which describes the relationship between the phase disturbance produced by a current impulse and the time at which the impulse is injected. For example a current impulse injected at the peak of oscillation has much less effect on the phase than an impulse injected at a zero-crossing. Their work has been used by Andreani et al. to develop closed-form accurate expressions for some of the commonly used LC tank oscillators [82–84].

The design community still has an unfinished journey to the realm of phase noise studies and new models and analyses are presented every year to advance the understanding of phase noise in electrical oscillators [85, 86].

4.3.4   Circuit Design and Layout

The VCO of Fig. 4.43c is implemented in 65 nm LP baseline CMOS technology. The cross-coupled pairs M1–M2 and M3–M4 generate the required negative conductance. The coupling transistors M5–M8 inject the output signals of one cross-coupled pair to the input of the other to produce anti-phase coupling required for quadrature generation. Since these transistors add extra parasitic capacitance to the oscillation nodes, they should preferably be small. However, by making them too small, the injection becomes weaker and the risk of frequency mismatch between the two tanks is increased. Therefore the dimension of the coupling transistors must be optimized to provide sufficient injection while minimally degrading the tuning range. After repetitive post-layout simulations the optimum size of the coupling transistors is found to be 25% of that of the cross-coupled ones. As mentioned before, the achievable tuning range is limited by the poor quality factor of the varactors at 60 GHz. Large-size varactors necessary for a large tuning range, lead to a low quality factor of the tank, higher phase noise and lower loop gain margin for oscillation. In this design, the varactors have a large number of short fingers to minimize the gate resistance and increase the quality factor. Big transistors are used as the current sources (Ibias) to lower their flicker noise and diminish the phase noise resulting from its upconversion. Source-follower buffers

Fig. 4.47 Source-follower
buffer used at the output

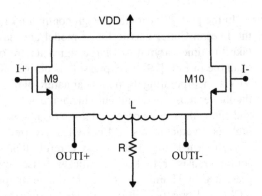

with center-tapped inductive load are used, as shown in Fig. 4.47. Minimum-size transistors are used to avoid extra parasitic capacitances which can deteriorate the tuning range. The resistor in the common source determines the biasing current of the differential pair, circumventing the need for current-mirror biasing of the gate of the transistors. Therefore, the gate of M9 and M10 can be connected directly to the output of the VCO, avoiding large RF-coupling capacitors.

Minimizing the parasitic capacitances at the output of the VCO core and maintaining the symmetry between I and Q parts are two major considerations made during the layout. An optimal trade-off has been made in the width of interconnects between transistors, buffers and the tank to achieve wide enough interconnects required for a high quality factor while keeping the additional parasitic capacitances in the oscillation node as small as possible. Special attention is paid to the symmetry of the layout to minimize the phase mismatch between I and Q outputs. The core transistors including cross-coupled pairs and injection transistors are arranged in a symmetrical ring configuration. While providing very good symmetry for the connections, this arrangement suffers from a mismatch in the orientation of the transistors and varactors, as shown in Fig. 4.48. On the other hand, arranging the transistors in an orientation-matched configuration, would introduce a length mismatch in the interconnects which can highly degrade the phase mismatch. Therefore, the former approach has been selected in this work. The coupling between inductors of the two tanks is another source of phase mismatch. Any DC path between the inductors might have unpredictable (or difficult-to-predict) impact on the mutual coupling and thus is avoided.

4.3.5 Measurement and Simulation Results

The measurement is performed on-wafer with high-frequency GSGSG Infinity probes. The measured tuning range is 5.6 GHz spanning from 57.5 to 63.1 GHz, as shown in Fig. 4.49a. According to Fig. 4.49b there is about 1 GHz difference in the simulated and measured tuning range. The effective tuning voltage across each

Fig. 4.48 Comparison between two different layout strategies: (**a**) the chosen strategy with symmetric interconnects and transistor orientation mismatch, (**b**) the strategy with matched transistor orientation and interconnect mismatch. D stands for drain and G stands for gate

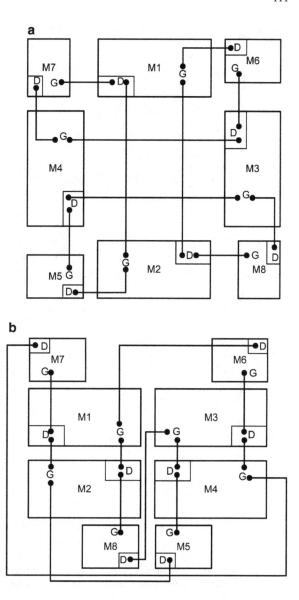

varactor varies between −0.6 and 0.6 V. In the design phase, the varactors were biased to 1.2 V resulting in a simulated tuning range of 4.4 GHz by varying the tuning voltage from 0 to 1.2 V. This tuning range was reduced to less than 3 GHz in the measurement. However a sharper variation in oscillation frequency is observed when the tuning voltage is varied between 0.6 and 1.8 V. For translating this range to 0–1.2 V, the VCO has to be redesigned with one end of the varactors biased to 0.6 V. The measured output power is also plotted in Fig. 4.49a for different biasing currents versus the tuning voltage. Higher biasing currents yield better output

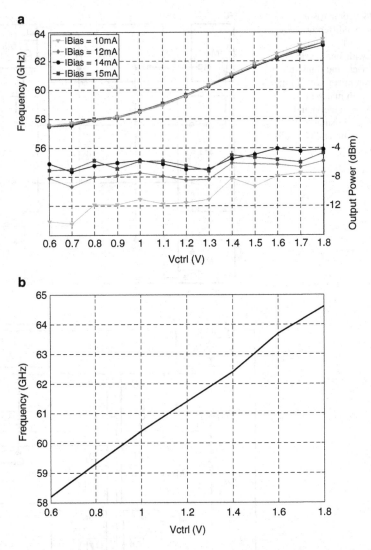

Fig. 4.49 (a) Measured oscillation frequency and output power versus tuning voltage for different biasing currents, (b) simulated oscillation frequency versus tuning voltage

power. The output power for IBias values of 14 mA and 15 mA varies from −7 to −4 dBm across the entire tuning range. The measured phase noise is −95.3 dBc/Hz at 1 MHz offset from the carrier with the lowest frequency, as shown in Fig. 4.50. The unusual shape of the phase noise curve at frequencies below 100 kHz originates from the internal PLL of the measurement equipment which is needed for stabilizing the oscillation frequency of the drifting VCO. As a result of high loop gain of the internal PLL the measured phase noise is suppressed in frequencies

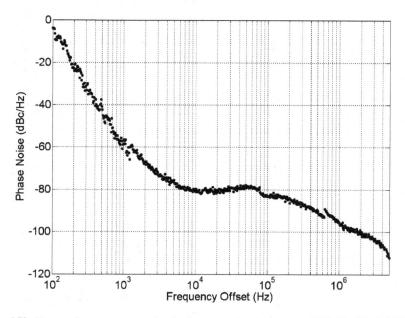

Fig. 4.50 Phase noise measurement for the lowest carrier frequency: −95.3 dBc/Hz @ 1 MHz offset from a 57.5 GHz carrier

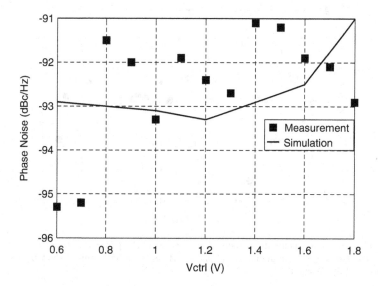

Fig. 4.51 Phase noise at 1 MHz offset from different carriers

close to the carrier. The phase noise at 1 MHz offset from different carrier frequencies is shown in Fig. 4.51. The measured power at the two I and Q outputs do not show any meaningful difference. The measurement of the phase imbalance

a

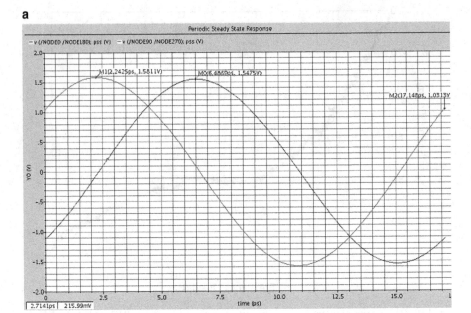

b

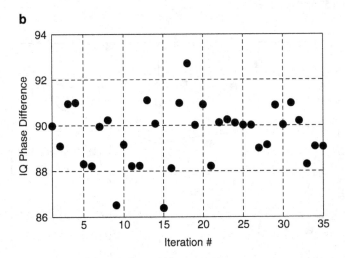

Fig. 4.52 Simulated quadrature outputs: (**a**) showing 89.1° phase difference for a nominal sample, (**b**) Monte Carlo simulation of IQ phase difference on 35 iterations showing a standard deviation of 1.38° from 90°

was not feasible due to the high frequency of the outputs and the enormous sensitivity of the phase mismatch to the length of the cables and interconnects used in the measurements. However the simulated results are shown in Fig. 4.52. Each VCO core, including a cross-coupled pair and an LC-tank, draws 15 mA from the 1.2 V supply. The two inductively-loaded source-follower buffers, used at the two outputs of the VCO, consume 20 mA in total.

State-of-the-art VCOs have been chosen for comparison in Table 4.3. This work exhibits a performance comparable to these VCOs. The formula used in calculation of figure of merit is obtained from:

$$FOM_{PN} = \pounds(\omega_0, \Delta\omega) - 20\log\left(\frac{\omega_0}{\Delta\omega}\right) + 10\log\left(\frac{P_{DC}}{1mW}\right) \qquad (4.71)$$

$$FOM_{FTR} = \pounds(\omega_0, \Delta\omega) - 20\log\left(\frac{\omega_0}{\Delta\omega} \bullet \frac{TR}{10}\right) + 10\log\left(\frac{P_{DC}}{1mW}\right) \qquad (4.72)$$

where $\pounds(\omega_0, \Delta\omega)$ is the single-sideband phase noise at a frequency offset $\Delta\omega$ from a carrier at ω_0, P_{DC} is the DC power consumption, and TR is the tuning range in percentage.

VCO and buffers occupy an area of 340 by 340 μm^2 as shown in the die photo of Fig. 4.53. Kim et al. [87] and [93] take advantage of SOI technologies with much lower parasitic capacitance compared to a bulk CMOS process to increase the tuning range. In Ref. [88] a very good phase noise is achieved by increasing the resonator quality factor, using an embedded artificial dielectric layer and very small varactors in the resonator. Using small varactors results in a small tuning range of 100 MHz. In Ref. [89] two VCOs are used to cover the entire ISM band and a complete phase-locked loop is demonstrated. The reason for the high power consumption in our work is that for compensating the loss of the low-Q varactors the negative conductance of the cross-coupled pair has to be increased which results in higher biasing current and hence higher power consumption.

4.4 Miller-Effect-Based VCO

Varactors are typically used for the realization of frequency tuning in conventional voltage controlled oscillators. Operating at high frequencies in conjunction with the non-ideal interconnects degrades the quality factor of the CMOS varactors which in turn confines the tuning range of the VCO. In practice the varactors dominate the finite quality factor of the resonator at high frequencies. The C_{max}/C_{min} ratio of the variable CMOS varactors are simulated to be as low as 2.1 at 60 GHz using a CMOS 65 nm technology. As a consequence achieving the required tuning range using these varactors is very difficult at 60 GHz.

As an alternative to the conventional approach, we present a tuning architecture based on the miller capacitance principle with a target tuning range of 10% and a target phase noise performance of -85 dBc/Hz [102]. The Miller effect tuning was first proposed in Ref. [103].

Table 4.3 Measurement results of the VCO compared with previous works

VCO [Ref]	Process Tech.	F_{osc} (GHz)	FTR (%)	Phase noise (dBc/Hz)	P_{Diss} (mW)	FOM_{PN} (dBc/Hz)	FOM_T (dBc/Hz)	Quad.
[87] ISSCC 07	65 nm SOI	70.2	9.55	−106.14 @ 10 MHz	5.4	−175.74	−175.34	No
[88] ISSCC 06	90 nm CMOS	60	0.16	−100 @ 1 MHz	1.9	−192.77	−156.85	No
[89] ESSCIRC 10 (complete PLL)	40 nm CMOS	66.5	10.5	−85 @ 1 MHz	66	−163.3	−163.7	Yes (mismatch not reported)
[90] RFIC 08	65 nm CMOS	54.1	11.5	−118 @ 10 MHz	7.2	−184	−185.3	No
[91] ASSC 10 (A 20 GHz PLL + Tripler)	65 nm CMOS	60.5	8.3	−96 @ 1 MHz	77.5	−172.7	−171.1	Yes (mismatch not reported)
[92] BCTM 08 (A single-phase VCO + polyphase filter)	0.25 μm SiGe	60.5	3.3	−92 @ 1 MHz	150	−165.9	−156.2	Yes (mismatch measured: 4.2°)
[93] IMS 04	90 nm SOI	56.5	14.7	−92 @ 1 MHz	21	−173.81	−177.16	No
This work	65 nm CMOS	60.29	9.3	−92.5 @ 1 MHz	36	−172.1	−171.5	Yes (mismatch simulated: 1.4°)

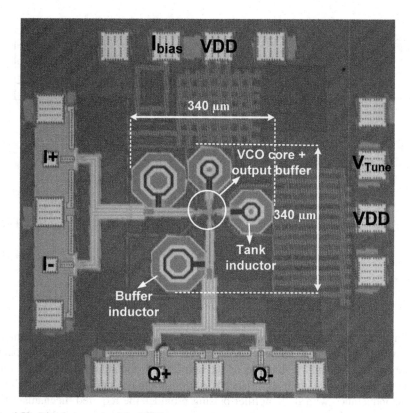

Fig. 4.53 Die photograph of the VCO

4.4.1   Tuning Using Miller Effect

Figure 4.54 illustrates the major principle behind the Miller effect, whereby the voltage gain is given by (4.73) and (4.74) and the effective input admittance is given by (4.75) and (4.76).

$$\frac{V_o}{V_i} = \frac{g_m Z_o + s Z_o C}{1 + s Z_o C}$$
(4.73)

$$\frac{V_o}{V_i} = B \angle \varphi$$
(4.74)

The quality factor (Q) of the effective input admittance (Yi) is given by (4.77). These parameters are a function of the amplifiers transconductance (g_m), the load impedance (Z_o) and the capacitance (C). They can also be expressed as a function of the magnitude of the voltage gain B and the phase of the overall voltage gain φ.

Fig. 4.54 Circuit-level
representation of the Miller
effect

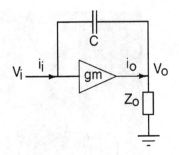

$$Y_i = j\omega C(1 - B\cos(\varphi) - jB\sin(\varphi)) \tag{4.75}$$

$$C_i = C(1 - B\cos(\varphi)) \tag{4.76}$$

$$Q = \frac{1 - B\cos(\varphi)}{B\sin(\varphi)} \tag{4.77}$$

Further inspection of these relations reveals that when the phase characteristic of the overall voltage gain approaches 180°, the value of the effective input capacitance becomes mainly a function of the amplifier transconductance and the real value of the effective input admittance becomes minimal. Controlling the amplifier transconductance with the amplifier biasing provides the required mechanism for the control of the effective input capacitance (C_i). The reduction of the effective input loss can be used to improve the quality factor of the effective input capacitance.

4.4.2 Design

The well known cross-coupled topology is used for the VCO core. Optimizing the tuning range for a certain frequency range in conjunction with fulfilling the oscillation criteria can be considered as the main target of this design paradigm. The phase noise performance was only a secondary design criterion. Accomplishing this target demands an appropriate overview of the parasitic capacitances of the active and passive components and non-ideal interconnections. A proper understanding of the relations between design parameters is also necessary.

Increasing the VCO core inductance will increase the quality factor and thus phase noise performance and increase the open loop gain, but it will decrease the capacitance value and thus the tuning range. Increasing the width of the VCO core cross-coupled transistors will increase the negative conductance and thus the open loop gain but it will also increase the parasitic capacitances and decreases the tuning range. To satisfy the oscillation criteria the open loop gain is chosen to be $\alpha = 3$ and the LC tank is tuned to the appropriate frequency.

Equations (4.78) and (4.79) describe general characteristics of an LC network containing a variable capacitance. The total capacitance, including the added variable capacitance C_{var} and the parasitics C_{par} is given by C_{tot}. L is the core inductor. Using a proprierty inductor model a quality factor (Q_L) of 30 can be assumed for the inductor. The quality factor of the variable capacitor (Q_C) is simulated to be 6.5. The overall LC tank quality factor is given by Q_{tank}.

$$C_{tot} = \frac{1}{\omega_0^2 L} \tag{4.78}$$

$$C_{var} = C_{tot} - C_{par} \tag{4.79}$$

Equation (4.80) relates the effective loss of the LC network (G_p) to the unloaded quality factor of the LC network. From these relations, one can conclude the minimum negative conductance ($G_{neg,min}$) of the cross-coupled oscillator. The above relation gives a lower boundary for the negative conductance. Figure 4.55a, b illustrate the attainable negative conductance as function of the width of the transistor and the inductance value. The attainable negative conductance has to be above the lower boundary for oscillation to take place. The frequency of operation dictates a direct relation between the inductor and the effective capacitance which is a function of parasitic capacitances of passive and active components and tuneability of the variable capacitance. Figure 4.55c illustrates the realizable minimum value of the capacitance for operating at the upper boundary of the required tuning range as function of the width of transistor and the value of the inductor. A combination of these two boundaries draws the general guideline for our design procedure. Figure 4.55d distinguishes realizable and infeasible areas as a function of transistor width and inductor value. Unachievable areas are presented at the upper right corner and the lower left corner because infeasible value for the minimum negative conductance and minimal variable capacitance. Projecting the required ratio between the maximum and minimum capacitance (C^{+var}/C^{-var}) into this picture will define the boundary for a proper trade-off between the power consumption and reducing the susceptibility to process variation. Figure 4.55d illustrates that a transistor width of 15 μm and an inductor value around 120 pH can be used as the initial values for design procedure.

$$G_P = \sqrt{\frac{C_{var}}{L}} \frac{Q_C + Q_L}{Q_C Q_L} \tag{4.80}$$

$$G_P = \sqrt{\frac{C_{var}}{L}} \frac{Q_C + Q_L}{Q_C Q_L} \tag{4.81}$$

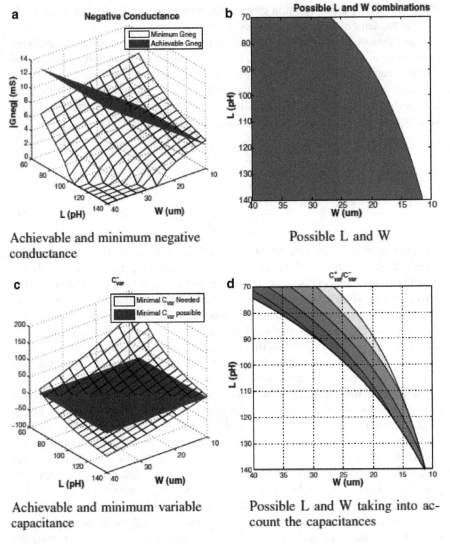

a Achievable and minimum negative conductance

b Possible L and W

c Achievable and minimum variable capacitance

d Possible L and W taking into account the capacitances

Fig. 4.55 Achievable solutions

Figure 4.56 illustrates the tuning circuit of the VCO. The effective capacitances at the nodes V_i^+ and V_i^- are used to tune the circuit, these capacitances are a function of the tail current Iss. Iss manipulates the gain of M3 and M4 and the gate-drain capacitances, Cg_d. Although the mirror current provides a robust and compact gain control system, its contribution to the deterioration of the phase noise performance has reduced its usefulness. At worst the current mirror noise contributed approximately 60% to the overall noise. The quality factor of the effective capacitance is a function of the voltage swing over the load which can deteriorate the phase noise performance. Although the implementation of a capacitance at the gate

Fig. 4.56 Miller capacitance circuit

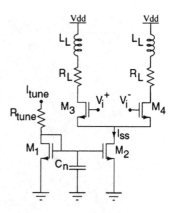

of transistors M1 and M2 can improve the phase noise performance, its practical value limits its effectiveness. Improving the quality factor of the effective capacitances is accomplished through the insertion of an inductor L_L, see Fig. 4.56. This inductor creates a low frequency zero at the left half plane and it contributes to the realization of 180° phase shift over entire tuning range.

4.5   Measurement and Simulation Results

For test and debugging purposes additional test structures, e.g. an inductor and an active capacitance have been taped out alongside the VCO (see Fig. 4.57). The inductor occupies the biggest area in the active Miller capacitor layout. Measurements have been made using an Agilent PNA Network Analyzer E8361A and an Agilent PSA Spectrum Analyzer E4446A extended with an Agilent preselected millimeter mixer 11,974 V. Measurement results are compensated for the external loss of cables and probes.

The simulation results have indicated a value of 134 pH with a quality factor of 20 for the VCO core inductors L which closely matches the measurements results of approximately 125 pH. Figure 4.58 depicts the simulation and measurement results of the active Miller capacitance. The measurement results are approximately 10fF lower than simulated which probably is due to overestimation of transistor parasitics. The simulation results indicated a value of 6 for the quality factor of the active Miller capacitance, although the measurement results point toward a Q value of 4.3.

Figure 4.59 shows the simulated and measured oscillation frequency of three different samples. Discrepancies between three samples are small. The simulated tuning range was 8%, the measurement showed 10.5% tuning range. This can be explained by the smaller than expected value for the Miller capacitance. A value of −13 dBm has been measured for the output power-level which is 2 dB lower than simulated. This difference can be caused by mismatch in the differential output paths and larger loss in the transmission lines. A varying shift of the center frequency in the range of 5 MHz, complicated a proper phase noise measurement.

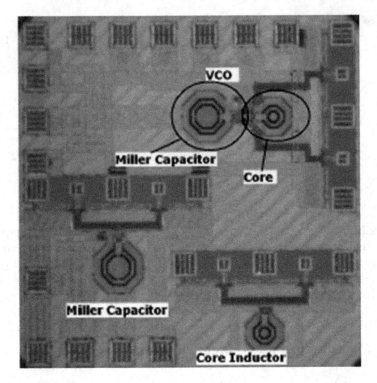

Fig. 4.57 Die photo

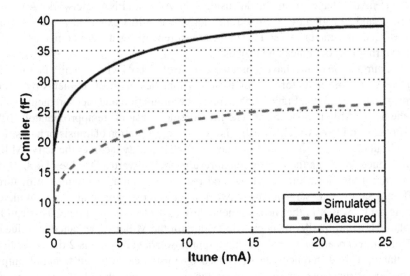

Fig. 4.58 Measured and simulated Miller capacitance

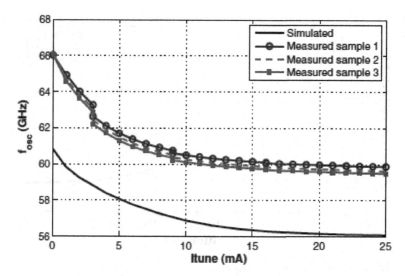

Fig. 4.59 Measured and simulated oscillation frequency

This behavior caused a shift of the operational frequency of the VCO during the phase noise measurement which made the measurement less reliable. The measured phase noise is shown in Fig. 4.60 as a function of the tuning current. The peak in phase noise is caused by the generated 1/f noise of the current mirror. From simulations it was estimated that the 1/f noise made up 60% of the total noise. At the high end of the tuning range, transistors M1 and M2 operate in the saturated region and they generate a modest amount of noise. At approximately 2 mA these transistors are biased just above threshold and the contribution of the 1/f noise will be maximal.

4.6 Conclusions

Three 60 GHz components are analyzed, designed, and measured. A two-stage fully integrated 60 GHz differential low noise amplifier is implemented in a CMOS 65 nm bulk technology. Utilizing a voltage–voltage feedback enables the neutralization of the Miller capacitance and the achievement of flat gain over the entire 6 GHz bandwidth. A noise and isolation analysis is done on the utilized topology.

An insightful analysis is performed on the white noise and conversion gain of voltage-driven switching-pair high-frequency active mixers. Unlike some previous works, the simple proposed analysis can predict the dependency of the mixer white noise on the local oscillator (LO) frequency. Besides, the noise contribution of the switching pair transistors is taken into account during the whole period of their operation and the assumption that the noise contribution is limited to the moment when both transistors are ON is discarded. This is verified by changing the time in which both transistors are ON and observing the effect on the simulated output

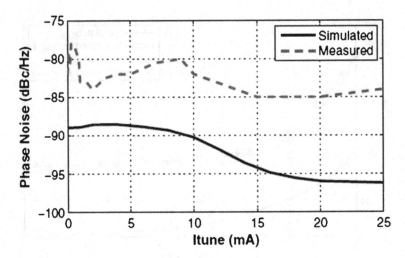

Fig. 4.60 Measured and simulated phase noise

noise. The relationships provided by the analysis, as verified by simulations, can reveal the influential factors on the gain and high-frequency noise and the potential approaches to improve the gain and noise performance.

An in-depth analysis is performed on the third-order intermodulation distortions (IMD3) in the switching pair of active CMOS mixers. The nonlinear time-varying switching pair is described by a hypothetical circuit composed of a nonlinear time-invariant circuit cascaded with a linear time-varying circuit. This allows us to apply time-invariant power and Volterra series for nonlinearity analysis. The analysis reveals that the impedance in the common-source node of the switching pair along with the fractions of the LO period in which both transistors are simultaneously ON are responsible for IMD3 generation. Besides it is noted that the IMD3 is composed of several terms which may in some cases cancel out each other. The analysis is validated with simulations.

A double-balanced 60 GHz mixer with ac-coupled RF input is designed and measured with a series capacitor in the input RF path to suppress the low frequency second order intermodulation distortions generated in the previous stage.

A quadrature 60 GHz VCO is presented which exhibits a comparable level of performance to state-of-the-art single-phase VCOs, despite the additional challenges and limitations imposed by the quadrature topology. Using varactors for frequency tuning, the VCO can cover a frequency range of 57.5 to 63.1 GHz. The tuning range is analyzed and is found to be limited by the poor quality factor of the varactors at high frequencies as well as the parasitic capacitances at the oscillation node.

The simulations are performed using the methodology described in Chap. 3. However, there are still discrepancies between simulation and measurement results. It is advisable to run an EM simulation on the whole layout in the future designs to reduce these discrepancies.

Chapter 5
Smart-Component Design at 60 GHz

In this chapter, circuit level solutions are presented for coping with the impact of process variations. One of the problems associated with process-variation-induced mismatch is the second order intermodulation distortion (IMD2). Based on the system-level analysis of Chap. 2, the overall performance of a receiver is more sensitive to the noise and nonlinearity distortion of the building blocks which contribute more to the total noise and distortion.[1] The contribution of the stages prior to the mixer in a zero-IF receiver to the total IMD2 can be suppressed by filtering. However, the IMD2 generated by the zero-IF mixer lies in the IF band and cannot be filtered. In Sect. 5.1, a tunable mixer is implemented for correcting the mismatches and minimizing the IMD2. Furthermore, since in many of the system-level analyses the IMD3 is presumed as the dominant source of intermodulation distortion, suppressing the IMD2 by this method prevents violating this assumption.

According to the system-level analysis of Chap. 2, accumulating the noise and nonlinearity contributions in one stage has the advantage of accumulating the sensitivities in that stage, i.e., the overall performance of the receiver becomes more sensitive to the noise and nonlinearity of that single stage and less sensitive to the noise and nonlinearity of the other stages. This way the performance degradations resulting from process variations can be compensated mostly by tuning the performance of that single stage. This idea is simulated at circuit level in Sect. 5.2.

---

[1] It is worth reminding that the component with highest noise (or highest nonlinearity) does not necessarily have the highest contribution to the total noise (or total nonlinearity distortion), because, as defined in Chap. 2, the noise or nonlinearity distortion contribution of a stage is also a function of the gain of its preceding stages.

P. Sakian et al., *RF-Frontend Design for Process-Variation-Tolerant Receivers,*
Analog Circuits and Signal Processing, DOI 10.1007/978-1-4614-2122-1_5,
© Springer Science+Business Media New York 2012

5.1 Wideband IP2 Correction

While offering the possibility of low-cost and compact solutions for receivers operating in the license-free band around 60 GHz, zero-IF receiver architecture suffers from problems such as dc offset and second order intermodulation distortions (IMD2). Since the IF bandwidth is around 1 GHz, the DC-offset can be prevented by high-pass filtering and losing a relatively small part of the IF bandwidth, or it can be solved by DC offset cancellation techniques. However, the IMD2 appears in the signal band and thus cannot be filtered. The interfering signals at that frequency range will be very common after successful commercial proliferation of 60 GHz transceivers. The intermodulation distortions resulting from these interferers and the nonlinearity of the receiver can significantly deteriorate the SNDR and bit error ratio.

The downconversion mixer is normally the main contributor to the IMD2. The low-frequency second-order distortions generated in the LNA or other preceding stages can be filtered by RF coupling or band-pass filtering. Using balanced topologies in the mixer design is beneficial in reducing the IMD2 by making it a common mode signal. However, unavoidable mismatches in the layout and the mismatches due to process variations can let some part of the common mode distortion to appear in the differential output. Due to the random nature of mismatches, it is not possible to obtain the required IMD2 by mismatch cancellation for all the fabricated chips during the circuit design phase. Therefore, sort of adaptability is required in the mixers to enable them for post-fabrication mismatch correction. In addition, multi-Gbps applications envisioned for 60 GHz band require around 1 GHz of IF bandwidth. Therefore, any IMD2 cancellation mechanism applied to a 60 GHz mixer must be functional across a wide frequency range. Thus, conventional narrowband IMD2 cancellation techniques are not beneficial. Figure 5.1 shows three different interferer scenarios with their resulting IMD2. In Fig. 5.1a the interferer consists of two closely-spaced tones resulting in a single-tone IMD2 at baseband which can be sufficiently suppressed by a narrowband mismatch cancellation technique. Figure 5.1b shows another scenario with a widely spaced three-tone interferer. Two of the tones are close to each other whereas the third tone is more distant. The resulting IMD2 tones are also far-off at the baseband spectrum. Therefore, a narrowband mismatch cancellation technique cannot suppress all the IMD2 tones simultaneously. Figure 5.1c shows a more general case in which the interferer is wideband just like the desired RF signal is. In this case, the resulting IMD2 is also a wideband spectrum rather than a set of single tones. Consequently, for suppressing the IMD2 in this case the mismatch cancellation must be operative in the whole IF band.

In this section a three-parameter tuning method is proposed and is shown both in theory and measurement to be effective in wideband cancellation of mixer IMD2. First, the conventional IMD2 cancellation methods are explained and the reasons of their ineffectiveness for wideband applications are clarified.

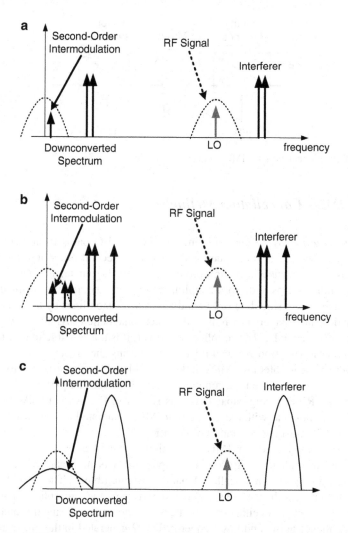

Fig. 5.1 Three different interferer scenarios with their resulting downconverted spectrum (the third and higher order distortions not shown): (**a**) closely spaced two-tone interferer, (**b**) widely spaced three-tone interferer, and (**c**) wideband interferer

Then, conditions for wideband functionality of IMD2 cancellation methods are derived and a three-dimensional method is proposed to tackle the issue. A circuit is proposed to implement and test the three-dimensional tuning method and the experimental and measurement results are presented.

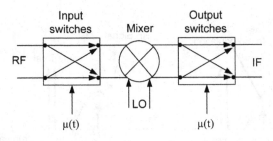

Fig. 5.2 Dynamic matching for IMD2 reduction in mixers

5.1.1   IMD2 Cancellation Methods

Careful symmetric layout and isolating the RF and LO paths to prevent RF-LO coupling are beneficial for reducing the level of second-order intermodulation distortions. For instance utilizing common-centroid layout techniques, including interdigitation patterns and cross-coupling can reduce the amount of mismatches. In addition applying shielding to LO and RF transmission lines by using neighboring grounded metal layers can reduce the electromagnetic crosstalk between them and hence reduce the RF-LO coupling. However, it is not possible to eliminate the random process-variation-induced mismatches during the layout.

To address the problem of IMD2 in the circuit-level, one approach is to improve the IMD2 produced by the transconductor [94]. However that part of IMD2 can be filtered by an RF-coupling capacitance, as shown in Fig. 4.17. In addition, these approaches do not deal with the problem of IMD2 generated by the switching pair, as a result of its mismatches and nonlinearities.

Another method used for IMD2 reduction is called dynamic matching [95]. The idea is illustrated in Fig. 5.2. The input signal is multiplied by a so-called mitigating signal $\mu(t)$ in a so-called dynamic matching block. Then it is fed to the mixer and afterwards the desired signal is restored by another multiplication with $\mu(t)$. The mitigating signal can either be periodic or pseudorandom. Using this method, the DC offset, flicker noise, and low frequency IMD2 generated in the mixer are either frequency translated or spread, depending on the chosen mitigating signal. Despite these appealing advantages, the dynamic matching method suffers from noticeable disadvantages. First of all, the circuit has to deal with additional interferers induced by $\mu(t)$ and its harmonics. In addition, generation of $\mu(t)$ and implemention of the dynamic matching blocks adds to the complexity of the system. Another disadvantage of this method is caused by the loss of the switches (especially at high frequencies) used in the dynamic matching blocks which can severely degrade the noise performance of the receiver. Lastly, the dynamic matching blocks also suffer from nonlinearity and mismatch which makes them potential contributors to the total IMD2.

Figure 5.3 illustrates an IMD2 cancellation technique which utilizes an extra path to regenerate the IMD2 and then deduct it from the IMD2 generated by the receiver [96]. Interestingly, the same concept is more efficiently implemented in

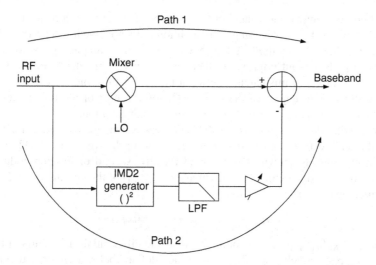

Fig. 5.3 Cancellation of IMD2 by generating it in a different path and subtracting it from the original

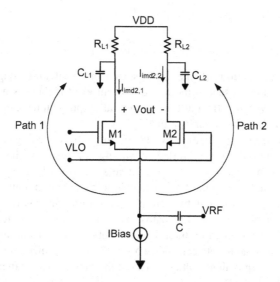

Fig. 5.4 The analogy between a single-balanced mixer and the IMD2 cancellation method of Fig. 5.3

balanced mixers, as shown in Fig. 5.4, because in Fig. 5.3 path 2 is only used to generate second order distortions and it does not improve the gain or other performance parameters, whereas in Fig. 5.4 both paths participate in the mixing process as well as generating second-order distortions which ideally add-up destructively in the differential output and cancel out each other. Therefore, in a balanced mixer, path 2 improves the conversion gain in addition to its positive impact on the IMD2 reduction. Thus, the method of Fig. 5.3 increases the complexity by adding an additional path without improving other performance parameters. However, both methods are prone to mismatches between the two paths which allow the second order distortions to appear at the output. Therefore, most of the commonly used IMD2 reduction methods are based on mismatch cancellation.

To explain different methods of IMD2 suppression which are based on mismatch cancellation, we first need to derive the relationship between the IMD2 and the mismatches. It was shown in Sect. 4.2.4 that the second order intermodulation distortions in the differential output voltage ($V_{imd2out}$) of the circuit of Fig. 5.4 can be written as below:

$$V_{imd2out} = 2I_{imd2CM}Z_L\delta Z_L + I_{imd2Diff}Z_L \tag{5.1}$$

where I_{imd2CM} $I_{imd2Diff}$ are the common-mode and differential-mode parts of $I_{imd2,1}$ and $I_{imd2,2}$, respectively. Z_{L1}, Z_{L2}, and δZ_L are defined below (with R_{out} defined as the resistance seen from the output node):

$$Z_{Li} = \frac{R_{outi}}{1 + R_{outi}C_{Li}j\omega} \quad (i = 1, 2) \tag{5.2}$$

$$Z_{L,1} = Z_L(1 + \delta Z_L)$$
$$Z_{L,2} = Z_L(1 - \delta Z_L) \tag{5.3}$$

I_{imd2CM} is a function of the input-stage and switching stage even-order nonlinearities and is present at the output current even if there is no mismatch in the circuit. However, its effect at the differential output voltage can be suppressed by a perfect matching between Z_{L1} and Z_{L2}.

$I_{imd2DIF}$ may be generated either by self-mixing or by switching pair nonlinearity combined with its mismatches [66]. Self-mixing is a result of the leakage of RF signal to the LO and vice versa. It is in general a function of the layout parameters and can be activated by any kind of mismatch in the LO or RF paths. The effect of self-mixing can be alleviated by careful and symmetric layout with sufficient isolation between the RF and LO paths. The impact of the switch nonlinearity and mismatch is determined by the mismatch between transistors in the switching pair [66]. The mismatch between the two transistors in a differential pair can be represented by an equivalent voltage offset at the gate of one of them [68]:

$$V_{off} = \Delta V_T + \frac{\partial f}{\partial \beta}\Delta\beta + \frac{\partial f}{\partial \theta}\Delta\theta$$

$$f = \frac{\theta}{\beta}I_{DS}\left(1 + \sqrt{1 + \frac{2\beta}{\theta^2 I_{DS}}}\right) + V_T \tag{5.4}$$

where f is an auxiliary variable with voltage dimension, I_{DS} is the biasing current of the transistor, $\beta = \mu_n C_{ox}W/L$, θ is a fitting factor taking into account the short channel effects, and V_T is the threshold voltage. Therefore, the effective mismatch between transistors can be controlled by inserting some intentional mismatch in the biasing voltages of the gates of transistors.

5.1.2 3D Tuning for Wideband IMD2 Cancellation

In this work it is assumed that the downconverted interfering signal at IF is wideband, whereas the interferer bandwidth is assumed small compared to its carrier frequency so that the interfering signal at RF can be considered as narrow-band. Thus wideband IMD2 cancellation means that the IMD2 must be suppressed over the entire IF bandwidth.

$V_{imd2out}$ can be minimized by tuning different parameters. Single-parameter tuning methods can adjust V_{off} to vary $I_{imd2DIF}$ [97]. They can also adjust output resistance mismatches (δR_{out}) or output capacitance mismatches (δC_L) to vary δz_L [98, 99]. To make the two terms in (5.1) cancel out each other by tuning only one parameter, the following should be satisfied:

$$\delta z_L = \frac{\delta R_{out} - R_{out} C_L \delta C_L j\omega_b}{(1 + R_{out} C_L j\omega_b)} = -\frac{I_{imd2Diff}}{2I_{imdCM}} \tag{5.5}$$

where ω_b is the IMD2 frequency at the output of the mixer. Higher powers of δR_{out} and δC_L are neglected in this approximation of δz_L. However, choosing only one parameter to tune, can only satisfy (5.5) at one single frequency point, because for each frequency the tunable parameter has a different optimum.

Even a two-dimensional tuning involving δR_{out} and δC_L is not sufficient [99], because it can only set (5.1) to zero at a single frequency point.

The approach chosen in this work is tuning all three parameters at the same time. This will result in simultaneous nullification of both terms in (5.1) as shown in (5.6) and (5.7). Since both δR_{out} and δC_L are set to zero in this approach, nullification of δz_L is frequency-independent. Due to narrowband assumption of interferer at RF, V_{off} can be chosen in a way that $I_{imd2Diff}$ is set to zero.

$$\delta z_L = \frac{\delta R_{out} - R_{out} C_L \delta C_L j\omega_b}{(1 + R_{out} C_L j\omega_b)} = 0 \tag{5.6}$$

$$I_{imd2Diff}\left(V_{off}\right) = 0 \tag{5.7}$$

5.1.3 Circuit Design and Layout

A double-balanced variant of the circuit of Fig. 5.4 is modified to implement the three-dimensional tuning explained in the previous section, resulting in a new

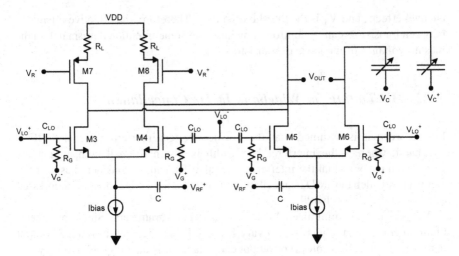

Fig. 5.5 Circuit schematic of the mixer with tunable output impedance and tunable gate biasing

circuit schematic as shown in Fig. 5.5. Variable resistors and varactors are added to the output, to provide tunability of the output impedance as required by (5.6). Variable resistors are in the simple form of series transistors biased in the triode region. The biasing of the gates of the switching pair transistors is adjusted separately for each half-circuit as required by (5.7).

The circuit is designed and fabricated in CMOS 45 nm technology. The supply voltage (VDD) is 1.1 V. V_R^- and V_R^+, which control the value of the variable resistors, are differentially tuned around 100 mV. V_C^- and V_C^+ control the varactors to tune the output capacitance and are differentially tuned around 500 mV. V_G^- and V_G^+ tune the biasing voltage of the gate of switching pair transistors and are differentially tuned around 0.9 V. The circuit draws less than 600 μA from VDD. The NMOS transistors M3–M6 are minimum-length and 6 μm wide with six fingers. The value of R_L is 580 Ω. The PMOS transistors M7-M8 are minimum-length and 4.5 μm wide with three fingers. The load varactors are 1.5 μm wide and 3.5 μm long. R_G is 490 Ω. The RF-coupling capacitances (C) are 98 fF each.

An IF-buffer, shown in Fig. 5.6, is used at the IF output to drive the 50 Ω load of the measurement equipment. The buffer is designed in a differential common-source configuration. The NMOS transistors are minimum-length and 32 μm wide with 32 fingers. The load resistances are 50 Ω each. Although this buffer is also prone to process-induced mismatch and hence can contribute to the total IMD2, it can be regarded as a part of the load and its mismatches can be corrected by tuning of V_R and V_C.

To reduce the number of required pads, two on-chip active baluns are used at the RF and LO paths. Therefore the single-ended-to-differential conversion is performed on-chip. The circuit used for RF and LO baluns is a differential common-source amplifier with inductive load as shown in Fig. 5.7. One input of the differential pair (the gate of M11) is only connected to a DC-biasing without any RF signal connected. The other input (the gate of M12) is not only DC-biased but also

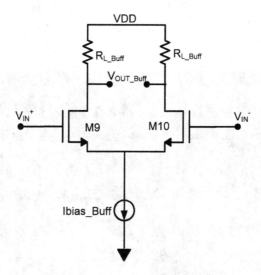

Fig. 5.6 Circuit schematic of the IF buffer used at the output of the mixer of Fig. 5.5

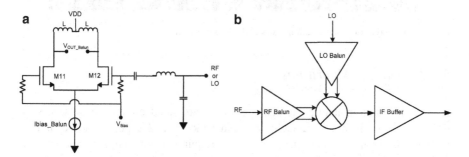

Fig. 5.7 (**a**) Circuit schematic of the LO and RF active baluns, (**b**) block diagram of the on-chip circuits

is connected to a matching network fed with LO or RF signal. The NMOS transistors M11–M12 are minimum-length and 16 μm wide with 16 fingers. The load inductor is a 240 pH center-tapped inductor with the center tap connected to VDD. The impact of these baluns on the IMD2 is expected to be negligible, since they are separated by RF-coupling capacitors from the mixer core.

The layout of the switching pairs is done in a symmetric ring configuration as shown in Fig. 5.8. The big devices which are partly seen on the right side are the RF coupling capacitors (C). A wider view of the layout showing more components is illustrated in Fig. 5.9 including the variable resistors, varactors, and IF buffer.

The complete chip, shown in Fig. 5.10, includes the mixer core, the two active baluns and matching networks at RF and LO inputs, and the IF buffer. Four inductors are used in the design. Two of them are used in the input matching networks and the other two are the loads of the active baluns. The circuit including the buffer and the baluns draws 21 mA from a 1.1 V source.

Current sources and   M3-M6
current mirrors
                                                    IF

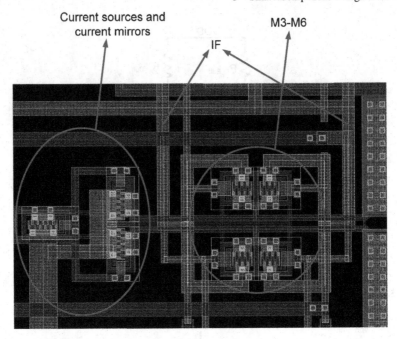

Fig. 5.8 The layout of the switching pairs and the current sources

5.1.4   Tuning Algorithms

Due to the random nature of mismatches in integrated circuits, it is necessary
to perform the mismatch cancellation, and hence IMD2 suppression, separately
for each fabricated chip. Furthermore, the required mismatch cancellation is a
function of the other parameters such as temperature and power supply variations.
Therefore, a calibration, preferably repeated in different time slots, is needed to
keep the IMD2 suppressed for every single chip during the whole operation time.
Figure 5.11 shows a calibration setup which has the potential of being implemented
on chip. In this work, the mixer (along with the baluns and buffers) is implemented
on-chip whereas the rest of the setup is realized in the measurement laboratory to
prove the concept. The first block on the left is a signal generator which generates
the interferer at the required frequency. Then the mixer downconverts the interferer
to the baseband resulting in the downconverted interferers, which fall out of IF
band, as well as second-order intermodulation distortions which fall in band.
The IMD2 measurement block measures the produced distortions and conveys
the information to the decision block. The decision block has to process the
information and demand the biasing controller to change the biasing of the gate
of the switching pair transistors, the varactors, and the variable resistors accordingly until a sufficiently low level of IMD2 is achieved. Therefore, the decision
block must take care of two tasks: processing the measured IMD2 to decide if it is
low enough and handling the biasing controller to adjust the mismatch.

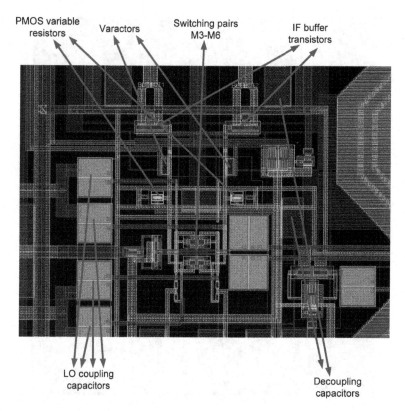

PMOS variable resistors   Varactors   Switching pairs M3-M6   IF buffer transistors

LO coupling capacitors   Decoupling capacitors

Fig. 5.9 The layout of the switching pairs, IF buffer, varactors, variable resistors, and decoupling and RF-coupling capacitors

Choosing the right algorithm to reach the optimum biasing for a suppressed IMD2 is a task that must be performed in the decision block. Since three parameters (V_R, V_C, and V_G) are to be tuned by the biasing controller for tuning the mixer of Fig. 5.5, the simplest possible tuning algorithm consists of covering the whole three-dimensional space formed by these three parameters with a high resolution to assure that the desired point is not missed. However, this algorithm is not time-efficient as it requires the maximum number of tuning steps to find the desired point with sufficiently suppressed IMD2. Alternatively, the decision block can stop searching as soon as the desired tuning point is achieved and thereby save the time considerably. However, in the worst case the number of tuning steps required for the latter is equal to the number of required tuning steps for the former as the desired point may reside at the end of the three-dimensional search space. The other alternative is to perform a low-resolution coarse tuning in the beginning and find the regions in which the IMD2 is the lowest. Then a fine tuning in the regions with the lowest IMD2 can locate the desired point. This algorithm is expected to reduce significantly the number of required tuning steps. Finally it is also possible to use more complicated algorithms such as gradient-based optimization to find

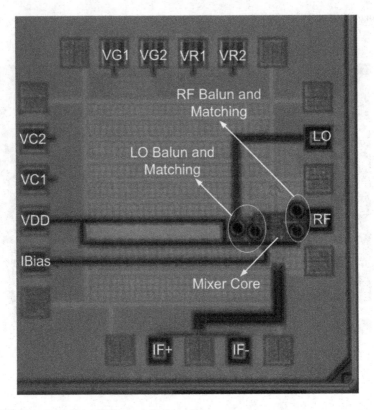

Fig. 5.10 Die photo of the IMD2 cancellation mixer

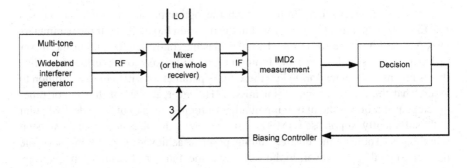

Fig. 5.11 The closed-loop configuration for IMD2 cancellation using the proposed mixer

the minima for IMD2 in a more time-efficient manner. In the next section the whole-space-covering algorithm and the coarse-tune fine-tune algorithm are compared in the course of the measurement in terms of their required number of steps.

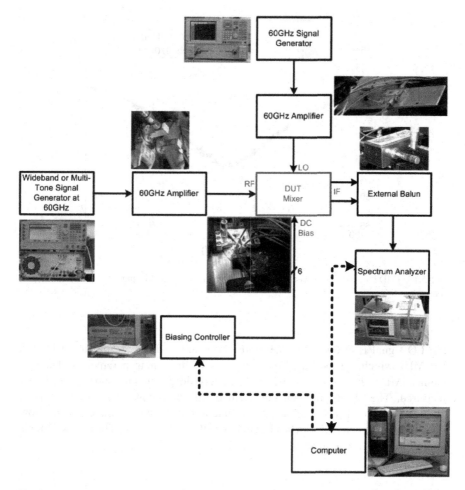

Fig. 5.12 The realization of the closed-loop configuration of Fig. 5.11 during the measurement for a multi-tone interferer

5.1.5   Measurement Results

The closed-loop configuration of Fig. 5.11 is realized with the measurement setup illustrated in Fig. 5.12. The spectrum analyzer is responsible for reading the IMD2 and conveying the information to a personal computer (PC) running Agilent VEE. The PC plays the role of the decision block of Fig. 5.11 and controls the biasing controller based on the measured IMD2.

To test the capability of IMD2 cancellation across a wide IF frequency range, a three-tone out-of-band signal is applied to the RF input of the mixer to emulate an out-of-band interferer. The three tones are at 61.070, 61.130, and 62.100 GHz.

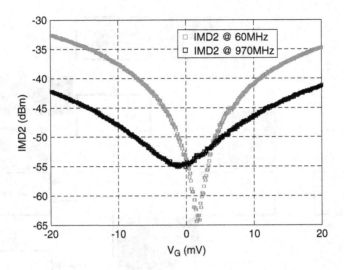

Fig. 5.13 IMD2 of sample 1 versus the mismatch in the gate biasing of the switching pair transistors

The LO signal is at 60 GHz. Therefore, the resulting IMD2 terms are at 60 and 970 MHz which are measured as a function of the tuning parameters. There is another IMD2 term at 1,030 MHz which is considered as out-of-band and is not measured. The closest fundamental term of the downconverted interferer is at 1,070 MHz which is also measured as a function of tuning parameters to see how much the conversion gain can be affected by IMD2 cancellation. The measurements are performed on several samples.

5.1.5.1 Single-Parameter Tuning

First of all, the single-parameter tuning is examined. Figures 5.13, 5.14, and 5.15 show the variation of IMD2 of one sample as a function of the control voltages of the gate of switching pair transistors, variable resistors, and varactors, respectively. The voltages are varied differentially around their corresponding common-mode values. It is observed that in single-parameter tuning the two IMD2 components are never simultaneously suppressed.

5.1.5.2 Two-Parameter Tuning

To verify the inadequacy of two-dimensional tuning, V_G is kept constant at zero while V_R and V_C are changed simultaneously. As shown in Fig. 5.16, V_R is swept from -40 to 40 mV in steps of 0.5 mV and in each step V_C is swept from

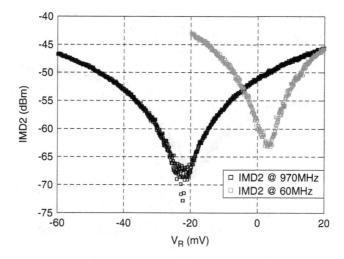

Fig. 5.14 IMD2 of sample 1 versus the variable resistance control voltage

−300 to 300 mV in steps of 5 mV. According to Fig. 5.16 the IMD2 at 60 and 970 MHz are never at the lowest points simultaneously, proving the inefficiency of two-dimensional tuning in this case.

5.1.5.3   Three-Parameter Tuning

As shown in Fig. 5.17 when V_G is also tuned an optimum point can be found where both 60 and 970 MHz IMD2 terms can be suppressed simultaneously. In this case three-dimensional tuning improves the IMD2 components at 60 and 970 MHz by 10 and 20 dBm respectively. Measuring IMD3 and fundamental components during the sweeps reveals that the maximum variations of these terms occur when V_G is varied. The IMD3 and the fundamental term are degraded by a maximum of 0.5 and 1 dB when V_G is varied by ±20 mV.

5.1.5.4   Single-Parameter Tuning on Other Samples

Measuring other samples reveals that three-dimensional tuning is not always required. As shown in Fig. 5.18, both IMD2 tones produced by another sample (referred to as sample 2 hereafter) can be suppressed by just tuning the gate biasing of the switching pair transistors. This can be understood by considering the random nature of mismatches in integrated circuits. If accidentally the mismatches in the transistors are dominant as compared to those of the load resistors and capacitors,

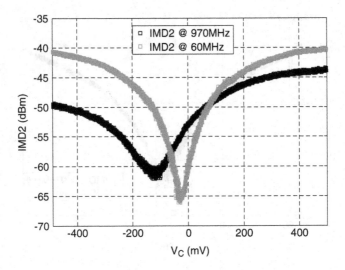

Fig. 5.15 IMD2 of sample 1 versus the varactor control voltage

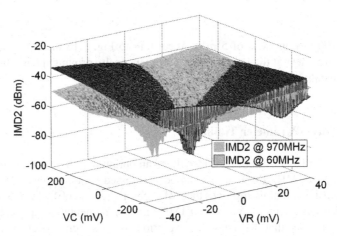

Fig. 5.16 IMD2 tuning of sample 1 by simultaneous variation of V_C and V_R while keeping V_G constant at 0: the two IMD2 tones are not suppressed simultaneously

the IMD2 is predominantly generated as a result of mismatch in the switching pair transistors. Therefore, in such a case tuning the mismatches of the switching pair is sufficient for suppressing the IMD2.

Figures 5.19, 5.20, and 5.21 show the results of single-parameter tuning for another sample (called sample 3 hereafter). They show the variation of IMD2 of sample 3 as a function of the control voltages of the gate of switching pair transistors, variable resistors, and varactors, respectively. It is once again observed that the two IMD2 components are never simultaneously suppressed with single-parameter tuning.

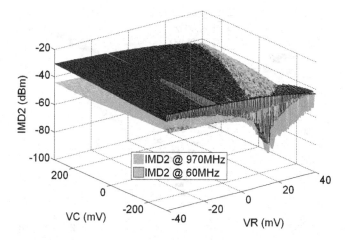

Fig. 5.17 IMD2 tuning of sample 1 by simultaneous variation of V_C and V_R while keeping V_G constant at -10 mV: the two IMD2 tones are suppressed simultaneously

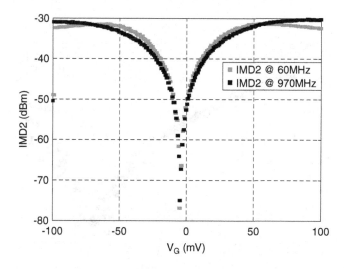

Fig. 5.18 IMD2 tuning of sample 2 by only changing V_G: the two IMD2 tones are suppressed simultaneously

5.1.5.5 Verification of Wideband Suppression of IMD2

Performing three-dimensional tuning on sample 3 suppresses both IMD2 tones at the same time, as shown in Fig. 5.22. The IMD2 tones before and after tuning, represented by markers 1 and 2, are shown in Fig. 5.22a, b, respectively. Both IMD2 tones are suppressed to close to the noise level after tuning. More suppression

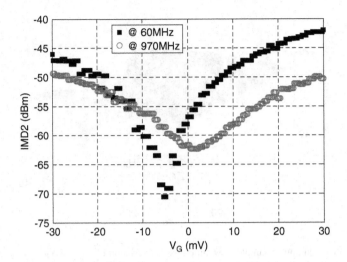

Fig. 5.19 IMD2 of sample 3 versus the mismatch in the gate biasing of the switching pair transistors

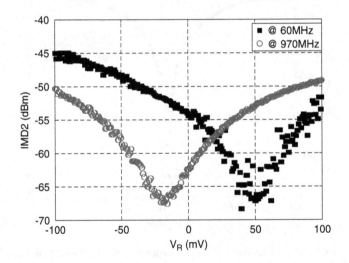

Fig. 5.20 IMD2 of sample 3 versus the variable resistance control voltage

requires a more accurate measurement with higher resolution bandwidth and a more accurate control with higher tuning resolution.

Higher resolution bandwidth and higher tuning resolution both demand a slower IMD2 cancellation process, because both measurement and tuning will take longer as the corresponding resolutions are increased.

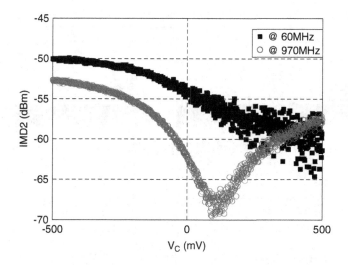

Fig. 5.21 IMD2 of sample 3 versus the varactor control voltage

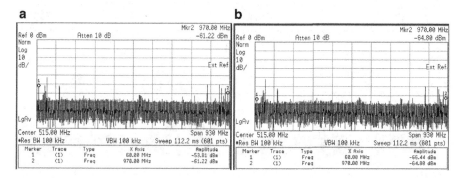

Fig. 5.22 IMD2 tones produced by sample 3 at 60 and 970 MHz (shown by markers *1* and *2*): (**a**) without tuning, (**b**) with tuning: suppressed to close to the noise level

To do an additional test on the capability of wideband IMD2 cancellation, the same tuning which results in the suppression of the two IMD2 tones at 60 and 970 MHz, as shown in Fig. 5.22, is preserved. Then the frequencies of the interfering tones are modified such that the resulting IMD2 tones fall at different frequencies. The measurement results, demonstrated in Fig. 5.23, show that these new IMD2 tones are kept suppressed without changing the tuning (values of V_G, V_R, and V_C) applied in Fig. 5.22. Therefore, according to this test, the required tuning for suppression of the IMD2 is independent of the frequency of the IMD2 tones. However, based on the assumptions of this work expressed in

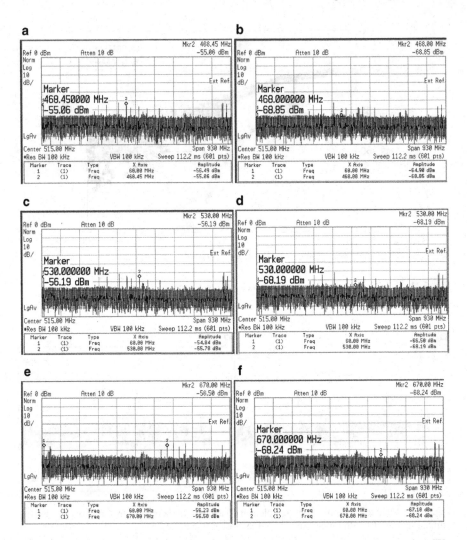

Fig. 5.23 Simultaneous suppression of IMD2 tones produced by sample 3 across the whole 1 GHz IF band (shown by markers *1* and *2*) at three exemplar points: (**a**) without tuning: unsuppressed IMD2 tones at 60 and 470 MHz, (**b**) preserving the same tuning as in Fig. 5.22b: suppressed IMD2 tones at 60 and 470 MHz, (**c**) without tuning: unsuppressed IMD2 tones at 60 and 530 MHz, (**d**) preserving the same tuning as in Fig. 5.22b: suppressed IMD2 tones at 60 and 530 MHz, (**e**) without tuning: unsuppressed IMD2 tones at 60, 670, and 730 MHz, (**f**) preserving the same tuning as in Fig. 5.22b: suppressed IMD2 tones at 60, 670, and 730 MHz

Sect. 5.1.2, this is only true as long as the frequency variations and differences are insignificant compared to the RF carrier frequency (while they can be significant compared to the IF as observed in the described test). For example, in this case the original three-tone interferer covers a frequency range of 1 GHz which compared to the LO frequency of 60 GHz comprises a relative bandwidth of 1.7%.

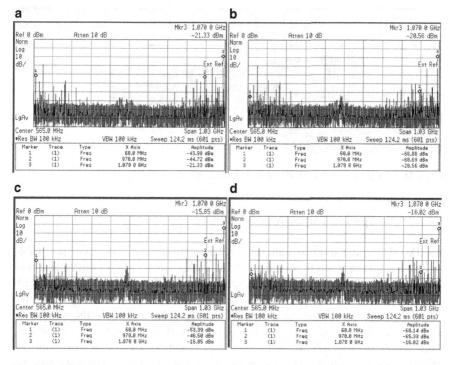

Fig. 5.24 IMD2 tones produced by sample 4 at 60 and 970 MHz (shown by markers *1* and *2*) and the downconverted interferer (shown by marker *3*): (**a**) without tuning and biasing current of 300 μA, (**b**) with tuning and biasing current of 300 μA, (**c**) without tuning and biasing current of 165 μA, (**d**) with tuning and biasing current of 165 μA

5.1.5.6 Impact of IMD2 Cancellation on Conversion Gain

To test the impact of IMD2 cancellation on the conversion gain, the IMD2 of another sample (called sample 4 hereafter) is measured and tuned while monitoring the amplitude of one of the downconverted interferer tones, as shown in Fig. 5.24. Markers 1 and 2 show the two IMD2 tones at 60 and 970 MHz and marker 3 shows one of the downconverted interferer tones at 1,070 MHz.

Figure 5.24a, b shows the unsuppressed IMD2 tones before tuning and the suppressed IMD2 tones after tuning, respectively, for a switching pair biasing current of 300 μA. Figure 5.24c, d shows the same for a switching pair biasing current of 165 μA. In the former case the IMD2 tone is suppressed by at least 16 dB whereas the conversion gain is degraded by less than 0.8 dB. In the latter case the IMD2 tone is suppressed by at least 15 dB whereas the conversion gain is degraded by only 0.2 dB.

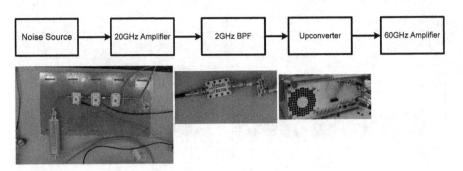

Fig. 5.25 The setup used for wideband interferer generation at 60 GHz

5.1.5.7 Comparing Tuning Algorithms

For IMD2 minimization of Sample 4 it takes 86,715 steps to cover the whole three-dimensional space by sweeping V_G mismatch from -15 to 14 mV in steps of 2 mV and in each step sweeping V_R mismatch from -40 to 40 mV in steps of 2 mV and in each step sweeping V_C mismatch from -300 to 400 mV in steps of 5 mV. Using a coarse-tune fine-tune algorithm can significantly reduce the number of required steps. First a coarse tuning is performed by sweeping V_G mismatch from -100 to 100 mV in steps of 40 mV and in each step sweeping V_R mismatch from -100 to 100 mV in steps of 40 mV and in each step sweeping V_C mismatch from -500 to 500 mV in steps of 200 mV. Then a fine tuning is performed in the area where both IMD2 terms are comparatively small, by sweeping V_G mismatch from -30 to 30 mV in steps of 10 mV and in each step sweeping V_R mismatch from 0 to 100 mV in steps of 20 mV and in each step sweeping V_C mismatch from -500 to 500 mV in steps of 100 mV. The total number of required steps in this case is 577. Therefore, by applying the coarse-tune fine-tune algorithm it is possible to reduce the number of steps from the worst case of 86,715 to just 577 in this case. Certainly, by applying more advanced algorithms it will be possible to achieve more reductions.

5.1.5.8 Measurement with Wideband Interferer

So far the tests have been performed using multi-tone interferers. To test the capability of the designed mixer in suppressing the IMD2 resulting from wideband interferers, the setup of Fig. 5.12 must be modified by replacing the multi-tone signal generator with a wideband signal generator. Due to unavailability of such a signal generator, the setup of Fig. 5.25 is used for generating a wideband signal at 60 GHz with an intended bandwidth of 2 GHz. A 20 GHz noise source is cascaded with some amplifiers and a band-pass filter and then fed to an upconverter.

   The signals generated by the setup of Fig. 5.25 at the output of the band-pass filter and at the output of the 60 GHz amplifier are shown in Fig. 5.26a, b, respectively.

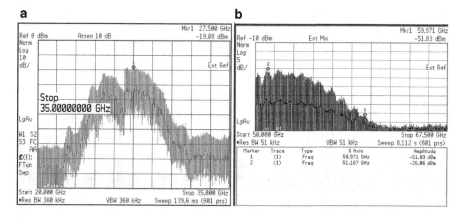

Fig. 5.26 Wideband signals generated by the setup of Fig. 5.25: (**a**) at the output of the band-pass filter and (**b**) at the output of the whole setup

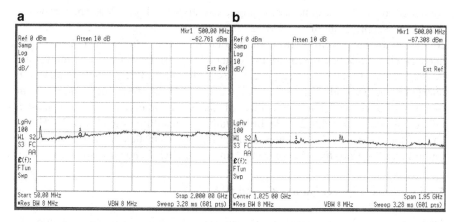

Fig. 5.27 The resulting IMD2 from the wideband interferer: (**a**) IMD2 with no tuning, (**b**) IMD2 after tuning

The bandwidth of the signal at the output of the band-pass filter is more than 4.5 GHz which is way above the intended 2 GHz bandwidth. The utilized upconverter suffers from a high RF-to-LO leakage which results in self-mixing and bandwidth spreading. Consequently the resulting wideband signal at 60 GHz has 7 GHz of bandwidth which is 3.5 times more than the intended 2 GHz.

Feeding the wideband signal generated in Fig. 5.26 to the setup of Fig. 5.12 results in the IMD2 shown in Fig. 5.27a. Performing the IMD2 cancellation flattens the spectrum of IMD2 as shown in Fig. 5.27b and reduces the power of IMD2 at 500 MHz by 5 dB. Preserving the same tuning and replacing the wideband interferer with a sinusoidal tone at 500 MHz results in the output shown in Fig. 5.28b which shows less than 0.8 dB degradation in conversion gain as

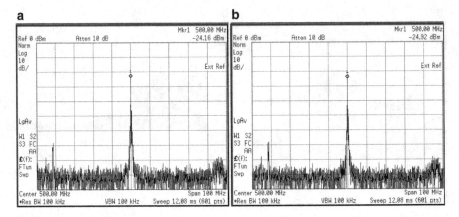

Fig. 5.28 The resulting output for a sinusoidal at 500 MHz (**a**) with no tuning, (**b**) with the tuning of Fig. 5.27b

compared to the case with no tuning applied as shown in Fig. 5.28a. The reason for the lower suppression of the IMD2 resulting from the wideband interferer as compared to that resulting from multi-tone interferer can be attributed to the unexpectedly high bandwidth of the generated interferer. In case of multi-tone interferer, the distance between the tones is around 1 GHz which renders the interferer narrowband with respect to the 60 GHz RF frequency. However in the case of wideband interferer, although the initial intention was to acquire a rather flat spectrum with 2 GHz bandwidth, the resulting signal has 7 GHz of bandwidth. Therefore, one of the assumptions made in this work regarding the limited bandwidth of the interferer compared to the RF frequency is violated.

5.1.5.9   Noise Figure, Gain, IP3, and IP2

The measured noise figure and power conversion gain of the mixer as a function of the IF frequency are shown in Figs. 5.29 and 5.30, respectively. The measured IIP3 and typical corrected IIP2 of the mixer are -7 and 27 dBm respectively.

The variation of gain with IP2 tuning was shown to be negligible in Fig. 5.24. The IP3 is observed to be most sensitive to V_G during the measurements and it varies by a maximum of 1 dB when V_G is changed from -20 to 20 mV. The noise figure variation as a result of IP2 tuning is found to be less than our measurement error which is about 0.5 dB.

5.1.6   Discussion

It is demonstrated both in theory and measurement that the presented three-dimensional tuning method can realize wideband cancellation of second order intermodulation distortion (IMD2) in a zero-IF downconverter. The resistance and

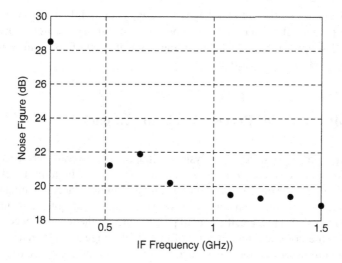

Fig. 5.29 The noise figure of the mixer versus IF frequency

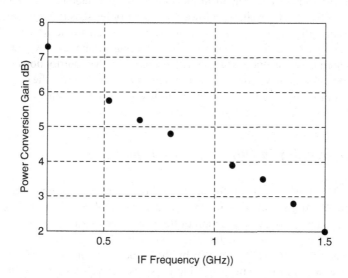

Fig. 5.30 The power conversion gain of the mixer versus IF frequency

capacitance at the output of the mixer as well as the gate biasing of the switching pairs are the three parameters used for tuning. A 60 GHz zero-IF mixer is designed and measured on-wafer to show that the proposed tuning mechanism can simultaneously suppress IMD2 tones across the whole 1 GHz IF band while having minor effect on the conversion gain and third order intermodulation distortion.

Repeating the tests for a wideband interferer with 7 GHz of bandwidth reveals that the amount of suppression is reduced for such wideband interferers. This is not unexpected as one of the assumptions behind this work is that the interferer is not wideband as compared to the RF carrier frequency (while it may be wideband with respect to the IF). Therefore, it can be a topic of further research to investigate methods of suppressing the IMD2 resulting from such interferers which are wideband compared to the RF carrier frequency.

5.2 Adjustment of Process Variation Impact

According to the system-level analysis of Chap. 2, the overall performance of a receiver is more sensitive to the noise and nonlinearity of building blocks which contribute more to the total noise and nonlinearity. Therefore, accumulating the noise and nonlinearity contributions in one stage has the advantage of accumulating the sensitivities in that stage, i.e., the overall performance of the receiver becomes more sensitive to the noise and nonlinearity of that single stage and less sensitive to the noise and nonlinearity of the other stages. This way the performance degradations resulting from process variations can be compensated mostly by tuning the performance of that single stage. The ideal location for such a stage is at the input of the receiver, i.e. the LNA, resulting in lowered second-order sensitivity to the gains of the stages. In addition, an LNA with tunable parameters can provide the possibility of input impedance corrections. However, implementing a tunable LNA is not a trivial task. Furthermore, it is not always easy to accumulate the contributions in the first stage. In the following we compare two strategies of accumulating the contributions in the LNA or in the mixer in terms of feasibility, costs, and benefits.

5.2.1 Tunability in Different LNA Topologies

Figure 5.31 shows three different LNA topologies: a common-gate LNA, an inductively degenerated common-source LNA, and a voltage-voltage transformer-feedback LNA. An inspection of these schematics suggests that the performance of these circuits is determined not only by the operating point and the parasitic of the transistor but also by the value of the inductors. Therefore, in general one needs to implement tunable passives, as well as variable biasing for transistors, to deal with the resulting performance degradations. For instance, the input-referred noise voltage of an inductively degenerated common-source LNA is derived from the following:

$$\bar{v}_{ni}^2 = \frac{4KT\gamma g_{d0}}{g_m^2} \frac{\left(R_S^2 + \left((L_G + L_S)\omega - \frac{1}{c_{gs}\omega}\right)^2\right)\left(\frac{g_m^2 L_S^2}{c_{gs}^2} + \left((L_G + L_S)\omega - \frac{1}{c_{gs}\omega}\right)^2\right)c_{gs}^2\omega^2}{\left((L_G + L_S)\omega - \frac{1}{c_{gs}\omega}\right)^2 + \left(\frac{g_m L_S}{c_{gs}} + R_S\right)^2}$$

$$(5.8)$$

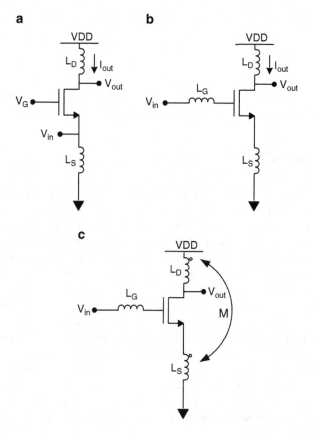

Fig. 5.31 (**a**) A common-gate LNA, (**b**) an inductively degenerated common-source LNA, and (**c**) a voltage-voltage transformer-feedback LNA

According to (5.8), correcting the performance degradations resulting from the variability of c_{gs} and g_m (optimistically not including L_G and L_S) requires using variable biasing and a varactor. Due to the low quality factor of varactors at high frequencies, the noise figure of the LNA would be degraded substantially by using varactors. If the inductors are also prone to variations due to process spreading, variable inductors turn into a necessity. Variable inductors can be implemented using active inductors [100]. However, such circuits fail to behave inductively at mm-wave frequencies. In addition, even at low frequencies they tend to deteriorate the noise performance due to the additional active circuitry. Using varactors in parallel with the inductors can also change the effective value of the inductance. However, as mentioned before, this method suffers from the low quality factor of the varactors at high frequencies which result in unacceptable levels of noise performance.

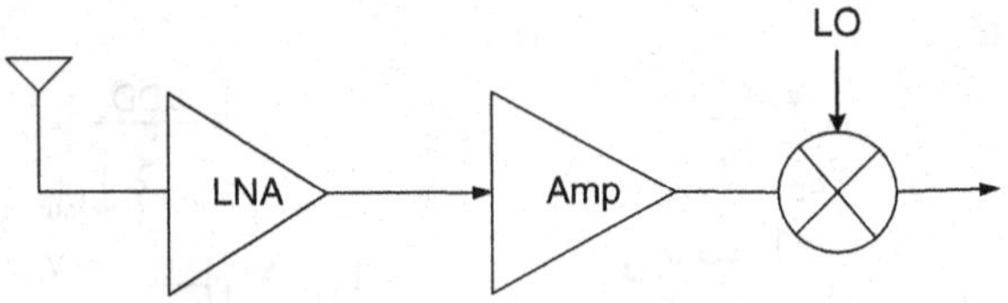

Fig. 5.32 Three-stage receiver used as the test vehicle in the study

5.2.2   Tunability in Other Stages

If for some reason, we fail in providing tunability to the LNA or rendering the LNA the dominant stage in terms of noise and nonlinearity contribution, it is still possible to design other stages, such as mixer, with such properties. In general, as explained in Chap. 2, the overall performance of the receiver is more sensitive to the performance of the stages with highest contribution to the total noise and nonlinearity. Therefore, if the noise and nonlinearity contribution of any stage is dominant compared to that of the other stages, it is practically possible to compensate the effect of process variations by just tuning the performance of that stage. Nevertheless, we know from Chap. 2 that the ideal situation is to accumulate all the contributions in the LNA and only if achieving this ideal becomes infeasible, we can opt for accumulation of the contribution, and hence the sensitivities, in other stages.

Furthermore, it is worth noting that achieving the best performance is the first priority for every designer. Therefore, the performance should never be compromised for accumulating the sensitivities in one stage. In other words, any design which provides such property and does not give the required performance is obviously not acceptable.

Choosing the mixer or another IF stage for tuning obviates the need of dealing with delicate high-frequency components. In addition as we will see later it is much easier to accumulate the noise and nonlinearity distortion contributions in the mixer.

5.2.3   Overall Design Considerations

In the previous subsections, we compared the difficulties of implementing tunability in the LNA, as a high frequency component, and mixer, as a circuit partly located at the low-frequency side. In this subsection we investigate the complexities of accumulating the contributions in the LNA and mixer. A three-stage direct-conversion 60 GHz receiver, shown in Fig. 5.32 is considered as a test vehicle for this study.

Two different design strategies are followed and compared. Ideally, as explained in Sect. 2.5, the ratio between the noise contribution of each stage to that of its following stage must be equal to the ratio between the distortion contribution of the stage to that of its following stage. This, along with an additional condition expressed by (2.35), assures a zero first order sensitivity of the overall performance to the gains of the

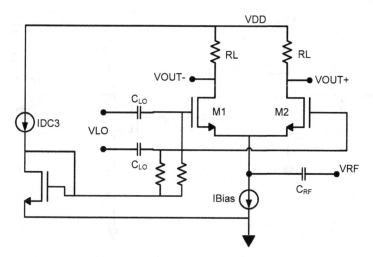

Fig. 5.33 Single-balanced mixer used in receiver 1 and receiver 2

individual stages. In the first strategy, represented by receiver 1, the noise and nonlinearity contributions are accumulated in the last stage (mixer). On the other hand, as explained in Sect. 2.5.2, to reduce the second order sensitivity of the overall performance to the individual gains, the noise and distortion contribution of the stages must be reduced in a step-wise manner as we move from the front stages to the rear ones. An attempt is made towards reaching this ideal situation in the second strategy. In the second strategy, represented by receiver 2, the focus is towards accumulating the noise and nonlinearity contributions in the first stage (LNA).

An inductively degenerated common-source LNA, shown in Fig. 5.31a, and a single-balanced mixer, shown in Fig. 5.33, are used in both designs as the first and third stage, respectively. Apart from different component parameters in the first and third stages of receiver 1 and receiver 2, the two receivers use two different circuit topologies as the second stage. A single-stage common-source amplifier is used as the second stage of receiver 1, whereas a two-stage amplifier consisting of an inductively degenerated common-source stage cascaded with a common-source stage is used as the second stage of receiver 2, as shown in Fig. 5.34. As a result, the second stage of receiver 2 provides a higher gain.

Figure 5.35 shows the noise and nonlinearity distortion contribution of each stage in receiver 1. According to Fig. 5.35b, the mixer has the highest contribution to the total nonlinearity distortion. Figure 5.35a shows that the mixer and LNA almost equally contribute to the total noise whereas the noise contribution made by the second stage is smaller. This means that the noise contribution and nonlinearity distortion contribution are mostly accumulated in the mixer. To shift this accumulation to the first stage, one way is to design a much more linear and low noise mixer, which is not realistic. Increasing the gain of the second stage can reduce the noise contribution of the mixer. However, it increases the nonlinearity distortion contribution of the mixer.

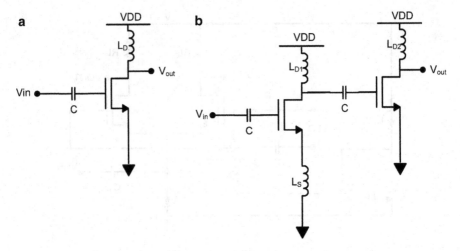

Fig. 5.34 The second stage in: (**a**) receiver 1, (**b**) receiver 2

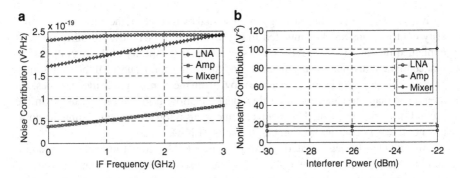

Fig. 5.35 Contribution of each component in receiver 1 to: (**a**) total noise, (**b**) total nonlinearity

Figure 5.36 shows the noise and nonlinearity distortion contribution of each stage in receiver 2. The noise contribution of the mixer is well diminished by increasing the gain of the second stage, while its nonlinearity distortion contribution is further increased. Therefore, only receiver 1 has the desired characteristic of accumulating the noise and nonlinearity distortion contribution in one stage. The main reason for not being able to accumulate the contributions in the LNA stems from the very high nonlinearity contribution of the mixer which cannot be reduced below the nonlinearity contribution of the LNA. However, this is not always true and the possibility of simultaneous accumulation of the noise and nonlinearity contribution in the LNA depends on many factors, such as the utilized circuit topologies, the application, the specifications, and the frequency of operation.

Based on Figs. 5.35 and 5.36, neither of the two receivers conforms to the aforementioned required conditions for zero sensitivity to the gains of individual stages. For instance, in receiver 1, the LNA and the mixer have the highest noise

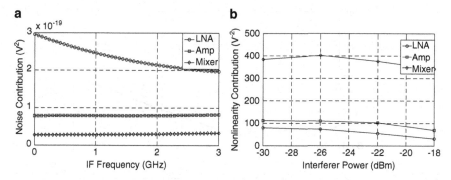

Fig. 5.36 Contribution of each component in receiver 2 to: (**a**) the total noise and (**b**) the total nonlinearity

Table 5.1 Simulation results of the two designs on process corners

	Receiver 1					Receiver 2				
	tt	ss	ff	sf	fs	tt	ss	ff	sf	fs
NF (dB)	6.54	6.61	6.45	6.46	6.47	5.55	5.6	5.45	5.55	5.56
IP3 (dBm)	−8.6	−8.1	−8.9	−13.5	−10.2	−15.8	−15.1	−15.8	−18.9	−17.5
NPD (dBm)	−74.1	−74.2	−74.1	−73.5	−74.1	−73.2	−73.6	−73.2	−69.8	−71.6
P_{DC}	12.3 mA × 1.2 V					18.4 mA × 1.2 V				

contribution while the LNA shows the lowest distortion contribution. Also, in receiver 2, the LNA has the highest noise contribution while it has the lowest distortion contribution. It appears that the combination of noise and nonlinearity for the blocks, as proposed by the system-level design for zero sensitivity to the gains, is not easily achievable at the circuit level. However, our attempt is to get as close as possible to the ideal situation sketched by the system-level design. Consequently, the desired property of accumulating the noise and nonlinearity distortion contribution in one stage, which is also proposed by the system-level method for facilitating the adaptability, is achieved in receiver 1.

The noise and nonlinearity performance of the two designs are simulated on the process corners and the results are listed in Table 5.1. The power consumption is kept constant on all process corners by keeping the biasing current constant. The simulations are performed with SPECTRE RF periodic steady state, periodic noise, and periodic ac analysis. The VDD is 1.2 V. The LO amplitude is 0dBm and the LO frequency is 60 GHz. In the next subsection, we investigate the possibility of correcting the performance of receiver 1 on its corners by just tuning the performance of the mixer, as the stage with accumulated noise and distortion contributions.

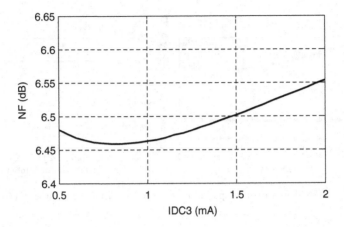

Fig. 5.37 Noise figure of receiver 1 at sf corner versus IDC3: NF is back to its typical value when IDC3 is 1.87 mA whereas the typical value of IDC3 is 1 mA

5.2.4 Correcting the Corner Performance

Now that we have a receiver with accumulated noise and nonlinearity distortion contribution in the mixer (receiver 1), the receiver performance degradations resulting from the process variations are expected to be (at least partly) correctable by just tuning the performance of the mixer. This idea is validated by further simulations performed on receiver 1. According to Table 5.1, the highest degradation of the IP3 of receiver 1 occurs at the slow–fast (sf) corner, while the NF is slightly improved at this corner. Choosing IDC3, used in the biasing of the mixer, as the tuning parameter, the corresponding performance variations are simulated. Figure 5.37 shows the variations of the receiver noise figure as a function of IDC3 at the slow–fast (sf) corner. The typical value of IDC3 is 1 mA. According to Fig. 5.37, the noise figure has its typical value when IDC3 is 1.87 mA. Then Fig. 5.38b shows that the IP3 is corrected to above its typical value when IDC3 is 1.87 mA.

Figure 5.39 shows the variations of the receiver noise figure as a function of IDC3 at the fast–slow (fs) corner. According to Fig. 5.39, the noise figure has its typical value when IDC3 is 1.5 mA. Then Fig. 5.40b shows that the IP3 is corrected to slightly less than its typical value when IDC3 is 1.5 mA.

As a conclusion, these tests validate the possibility of correcting the performance of the whole receiver on the process corners, by only tuning the performance of a single stage. This is possible, because the noise and distortion contributions are accumulated in that single stage (in this case the mixer). This possibility can greatly facilitate the correction of process-induced performance degradations in a smart receiver, as it confines the required number of tunable parameters.

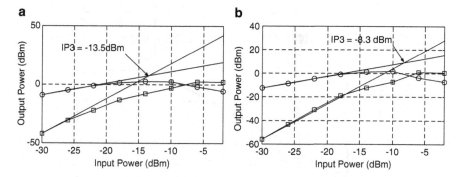

Fig. 5.38 IP3 of receiver 1 at sf corner for: (**a**) IDC3 of 1 mA and (**b**) IDC3 of 1.87 mA

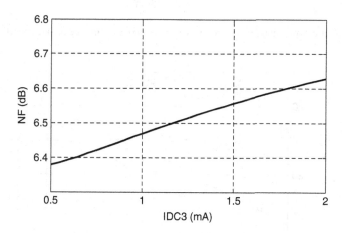

Fig. 5.39 Noise figure of receiver 1 at fs corner versus IDC3: NF is back to its typical value when IDC3 is 1.5 mA whereas the typical value of IDC3 is 1 mA

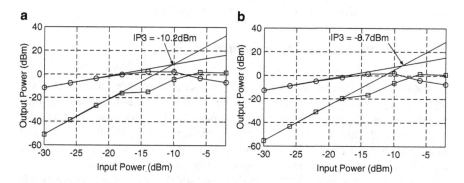

Fig. 5.40 IP3 of receiver 1 at fs corner for: (**a**) IDC3 of 1 mA and (**b**) IDC3 of 1.5 mA

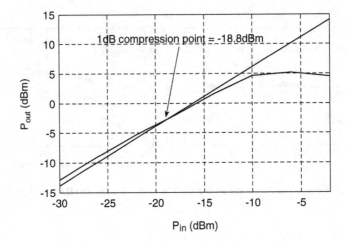

Fig. 5.41 1 dB compression point of receiver 1 after approximating the layout impact

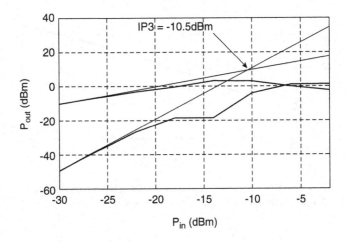

Fig. 5.42 IP3 of receiver 1 after approximating the layout impact

5.2.5   Layout Impact Approximated

To obtain a more accurate estimate of the receiver performance, the impact of the layout is approximated by adding more parasitic resistance and reactance at some specific points. The values of the additional parasitic are calculated based on prior experience with mm-wave layout and post-layout simulations. The RF lines, ground lines, and some of the DC lines are modeled by RLC π-networks. The resulting typical performance parameters are shown in Figs. 5.41, 5.42, and 5.43.

Comparing these results with Table 5.1, reveals that the noise performance is hardly degraded after adding the parasitic and the total IP3 is degraded by just 2 dB.

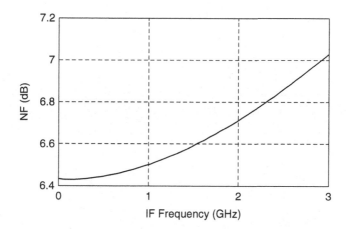

Fig. 5.43 NF of receiver 1 after approximating the layout impact

5.3 Conclusions

A tunable mixer is presented for minimizing the IMD2 across a wide IF bandwidth. It is demonstrated both in theory and measurement that the presented three-dimensional tuning method is beneficial for wideband cancellation of second order intermodulation distortion (IMD2) in a zero-IF downconverter. The resistance and capacitance at the output of the mixer as well as the gate biasing of the switching pairs are the three parameters used for tuning. A 60 GHz zero-IF mixer is designed and measured on-wafer to show that the proposed tuning mechanism can simultaneously suppress IMD2 tones across the whole 1 GHz IF band while having minor effect on the conversion gain and third order intermodulation distortions.

A receiver is designed with good noise and nonlinearity performance and with accumulated noise and nonlinearity distortion contribution in its last stage (mixer). As a result, the overall performance of the receiver is more sensitive to the performance variations of the mixer. Simulations show that it is possible to correct the overall receiver performance degradations resulting from process variations by just tuning the performance of the mixer. In fact, these simulation tests validate the possibility of correcting the performance of the whole receiver on the process corners, by only tuning the performance of a single stage (in this case the mixer and only one parameter of the mixer). This possibility can greatly facilitate the correction of process-induced performance degradations in a smart receiver, as it confines the required number of tunable parameters. It can also facilitate the performance trimming of the fabricated chips in the production line. The circuits are also simulated with additional parasitics, calculated based on previous design experience, to approximate the impact of interconnects added during the layout.

Chapter 6
Conclusions and Recommendations

In this chapter the conclusions of the book and also recommended future works are presented.

6.1 Conclusions

Design guidelines have been developed to cope with the impact of process variations on the performance of the receivers, high-performance circuit blocks have been designed for a 60 GHz receiver, and eventually reconfigurable circuits have been proposed to improve the robustness of the receiver to process variations with an emphasis on the simplicity and efficacy of the tuning by using a small set of tunable parameters.

A system-level analysis has been performed on a generic RF receiver. First, bit error rate (BER) was considered as a figure of merit representing the overall performance of the receiver. Then, each stage of the receiver was described by three parameters: voltage gain, noise, and nonlinearity which are prone to variation due to process spread. The variation of these parameters represents all lower-level sources of variability. Since bit error rate (BER), as a major performance measure of the receiver, is a direct function of the noise and distortion, the contribution of each block to the overall noise plus distortion (NPD) has been analyzed, opening the way for minimization of the sensitivity of the NPD to the performance variation of individual stages. It has been shown that the first order sensitivities of NPD to the individual gains of the building blocks can all be made zero. Its second order sensitivity to the gains of the building blocks can be reduced. Its sensitivity to noise and nonlinearity of an individual building block can be reduced, but at the cost of that of other blocks; thus, its sensitivity to noise and nonlinearity cannot be reduced over the whole system. Applying the analysis to a zero-IF three-stage 60 GHz receiver shows a significant improvement in the design yield, by nullifying the first order sensitivities of the overall performance to the individual gains of the blocks. Reduction of the second order sensitivity of the NPD to the gain of

P. Sakian et al., *RF-Frontend Design for Process-Variation-Tolerant Receivers*,
Analog Circuits and Signal Processing, DOI 10.1007/978-1-4614-2122-1_6,
© Springer Science+Business Media New York 2012

individual stages, by keeping the contribution factor of all the stages below one, results in further improvements.

After identifying the limitations of a pure system-level approach, i.e., inability to suppress the sensitivity of the overall performance to the noise and nonlinearity of all the blocks, the focus was shifted towards circuit-level methods by providing re-configurability to some RF circuits. A receiver has been designed with good noise and nonlinearity performance and with accumulated noise and nonlinearity distortion contribution in its last stage (mixer). As a result, the overall performance of the receiver is more sensitive to the performance variations of the mixer. Simulations show that it is possible to correct the overall receiver performance degradations resulting from process variations by just tuning the performance of the mixer. This can greatly facilitate the calibration of the receiver in a smart receiver or its performance trimming in the production line. Furthermore, a tunable mixer has been presented for minimizing the IMD2 across a wide IF bandwidth. It has been demonstrated both in theory and measurement that a presented three-dimensional tuning method is beneficial for wideband cancellation of second order intermodulation distortions (IMD2) in a zero-IF downconverter.

To address the challenges of 60 GHz circuit design, a design methodology has been utilized which serves to properly model the parasitic effects and improve the predictability of the performance. The parasitic effects due to layout, which are more influential at high frequencies, are taken into account by performing automatic RC extraction and manual L extraction. The long signal lines are modeled with distributed RLC networks. The problem of substrate losses is addressed by using patterned ground shields in inductors and transmission lines. The cross-talk issue is treated by using distributed meshed ground lines, decoupled DC lines, and grounded substrate contacts around sensitive RF components. However, in practice, it has been observed that accurate simulation of all the effects is sometimes very time consuming or even infeasible. For instance electromagnetic simulation of a transformer in the presence of all the dummy metals is beyond the computational capability of existing EM-simulators.

The on-wafer measurements on the 60 GHz circuits designed in this work have been performed using a waveguide-based measurement setup. The fixed waveguide structures, specially provided for the probe station, serve for the robustness of the setup as they circumvent the need for cables, which are by nature difficult to rigidify, in the vicinity of the probes. Taking advantage of magic-Ts, it is possible to measure differential mm-wave circuits with a two-port network analyzer rather than using a much more expensive four-port one. Noise, s-parameter, and phase noise measurements have been performed using the mentioned setups.

6.2 Recommendations

In the course of the research some interesting topics for future work have been identified.

It is advisable to do another case study based on the system-level design guidelines of Chap. 2 on a realized and measured receiver including the phases of the intermodulation distortion components of all the blocks to validate all the derived general-case sensitivity relationships with measurements.

Any pure system-level design methodology without regarding the circuit-level limitations can yield unrealistic specifications for the building blocks of the system. Therefore, using more circuit-level information during the system-level design can result in a more relevant outcome. Consequently, it is advisable to derive and use relationships between different performance parameters of the utilized circuit topologies and adopt them in the system design. For instance, it is well known that it is not possible to have any desired combination of noise figure, gain, and IP3 in a certain RF circuit. Therefore, if the space of all possible combinations is derived for the circuit, it is possible to reject all outcomes of the system-level design which call for impossible combinations of performance parameters. Alternatively, if the relationship between the noise figure, gain, and IP3 of each block is known based on the utilized circuit topology, these relationships can be used in the system-level analysis, leading to more realistic specifications.

The possibility of suppressing the second order intermodulation distortion (IMD2) arising from wideband interferers at 60 GHz has been demonstrated in this book. However, the interferer, although wideband as compared to the IF, was assumed to be relatively narrowband compared to the RF carrier frequency. Therefore, it can be a topic of further research to investigate methods of suppressing the IMD2 arising from interferers which are wideband even compared to the RF carrier frequency.

Appendix A
Optimizing Total Power Consumption by Defining Total Noise and Total Nonlinearity

Using (2.6) and considering third order nonlinearity as the only dominant order of nonlinearity, $V_{int,tot}^2$ can be described as a function of $V_{IP3i,tot}^2$:

$$\bar{V}_{ni,tot}^2 = \frac{NPD - kTB}{B(R_s + R_{in})^2} \times 4R_s R_{in}^2 - \frac{V_{interferer}^6}{BV_{IP3i,tot}^4} \qquad (A.1)$$

Replacing the result in (2.31), we have:

$$P_{tot} = \frac{BV_{IP3i,tot}^6}{\frac{NPD-kTB}{(R_s+R_{in})^2} \times 4R_s R_{in}^2 \times V_{IP3i,tot}^4 - V_{interferer}^6} \times \left(\sum_{m=1}^{N} \sqrt[3]{P_{Cm}} \right)^3 \qquad (A.2)$$

Taking the derivative of P_{tot} with respect to $V_{IP3i,tot}^2$, we have:

$$\frac{\partial P_{tot}}{\partial \left(V_{IP3i,tot}^2 \right)} = \frac{BV_{IP3i,tot}^4 \left(\frac{NPD-kTB}{(R_s+R_{in})^2} \times 4R_s R_{in}^2 \times V_{IP3i,tot}^4 - 3V_{interferer}^6 \right)}{\left(\frac{NPD-kTB}{(R_s+R_{in})^2} \times 4R_s R_{in}^2 \times V_{IP3i,tot}^4 - V_{interferer}^6 \right)^2} \times \left(\sum_{m=1}^{N} \sqrt[3]{P_{Cm}} \right)^3 \quad (A.3)$$

Setting (A.3) to zero, we have:

$$V_{IP3i,tot}^4 = \frac{3V_{interferer}^6}{\frac{NPD-kTB}{(R_s+R_{in})^2} \times 4R_s R_{in}^2} \qquad (A.4)$$

Using (2.3) and (2.5), $V_{IMD3i,tot}$ can be obtained:

$$V_{IMD3i,tot}^2 = \frac{NPD - kTB}{(R_s + R_{in})^2} \times 4R_s R_{in}^2 \times \frac{1}{3} = \frac{B\bar{V}_{ni,tot}^2}{2} \qquad (A.5)$$

which results in (2.21).

P. Sakian et al., *RF-Frontend Design for Process-Variation-Tolerant Receivers*,
Analog Circuits and Signal Processing, DOI 10.1007/978-1-4614-2122-1,
© Springer Science+Business Media New York 2012

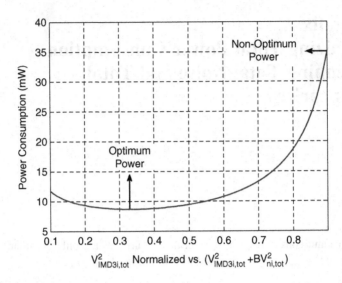

Fig. A.1 Power consumption given by the optimum-power design method as a function of $V^2_{IMD3i,tot}$ for a constant NPD.

Figure A.1 shows the power consumption of a receiver (described in detail in Sect. 2.8) as a function of $V^2_{IMD3i,tot}$. In other words, $V^2_{IMD3i,tot}$ and $BV^2_{ni,tot}$ are varied in a way that their sum remains constant so that NPD remains unchanged. For each value of $V^2_{IMD3i,tot}$ and $BV^2_{ni,tot}$ the power consumption is calculated according to optimum-power method using (2.31). Based on Fig. A.1, minimum power consumption is achieved when $V^2_{IMD3i,tot}$ constitutes one third of ($V^2_{IMD3i,tot} + BV^2_{ni,tot}$) which is in agreement with (2.21). In the extremes the power consumption can be much higher, for example when $V^2_{IMD3i,tot}$ comprises 90% of ($V^2_{IMD3i,tot} + BV^2_{ni,tot}$), the blocks need to be extremely low-noise and the power consumption can be more than four times higher than the minimum.

Glossary

Symbol	Description	Unit
A	Amplitude of the sinusoidal output of the VCO	V
A_v	Voltage conversion gain of the mixer	
$A_{vn,k}$	Voltage gain of the kth stage at the signal band including the loading effects	
$A_{vn,k,image}$	Voltage gain of the kth stage at the image band including the loading effects	
$A_{vn,tot}$	Voltage gain of the whole receiver at the signal band	
$A_{vn,tot,image}$	Voltage gain of the whole receiver at the image band	
B	Effective noise bandwidth	Hz
B_{CS}	B element of the T matrix of a transistor in common-source configuration	Ω
B_x	B element of the T matrix of a transistor in source-follower configuration	Ω
B_{sig}	Signal bandwidth	Hz
C_{max}	Maximum capacitance of the LC-tank	F
C_{min}	Minimum capacitance of the LC-tank	F
$C_{ImA,k}$	Ratio between the square of voltage gain at the image band to the square of voltage gain at the signal band calculated for the kth stage	
$C_{ImN,k}$	Ratio between the square of the noise voltage at the image band to the square of the noise voltage at the signal band calculated for the kth stage	
$C_{Distortion, q}(k)$	Contribution of the kth stage to the total q^{th} order intermodulation distortion	V^2
c_{gd}	Gate-drain capacitance	F
c_{gs}	Gate-source capacitance	F
C_L	Load capacitance	F
C_{L1}	The capacitance seen from the positive node of the differential output	F
C_{L2}	The capacitance seen from the negative node of the differential output	F
$C_{Noise}(k)$	Contribution of the kth stage to the total noise	V^2
C_{ox}	Oxide capacitance of a MOS transistor	F/m^2

(continued)

Symbol	Description	Unit
C_P	The equivalent capacitance after performing series-to-parallel conversion on the tank capacitance	F
C_s	Capacitances at the common source of the switching pair of the mixer	F
D_{CS}	D element of the T matrix of a transistor in common-source configuration	
D_x	D element of the T matrix of a transistor in source-follower configuration	
F	An empirical parameter called effective noise figure by Leeson	
G_1	First order Volterra kernel from RF current to the gate-source voltage of the mixer	Ω
G_2	Second order Volterra kernel from RF current to the gate-source voltage of the mixer	V/A^2
G_3	Second order Volterra kernel from RF current to the gate-source voltage of the mixer	V/A^3
g_{d0}	Transistor channel conductance when the drain-source voltage is zero	$1/\Omega$
g_m	First-order transconductance of the transistor	$1/\Omega$
g_{m2}	Second-order transconductance of the transistor	A/V^2
g_{m3}	Third-order transconductance of the transistor	A/V^3
G_C	Power conversion gain of the mixer	
G_T	Transducer power gain	
g_s	Inverse of R_S	$1/\Omega$
I_{DS}	Drain-source biasing current of the transistor	A
$I_{imd2,1}$	IMD2 current in the first branch of the differential output of the mixer	A
$I_{imd2,2}$	IMD2 current in the second branch of the differential output of the mixer	A
I_{imd2CM}	Common-mode output IMD2 current of the mixer	A
$I_{imd2Diff}$	Differential-mode output IMD2 current of the mixer	A
I_{Out}	Differential output current of the switching pair of the mixer	A
I_{Out3}	Third-order intermodulation content in the differential output current of the switching pair of the mixer	A
I_{pk}	The amplitude of the sinusoidal current	A
I_{RF}	RF current flowing into the switching pair of the mixer	A
IP3	Third order input intercept point	dBm
k	Boltzmann constant $= 1.38 \times 10^{-23}$	J/K
k_c	Coupling factor between two inductors	
L_D	Drain inductance	H
L_G	Gate inductance	H
L_P	The equivalent inductance after performing series-to-parallel conversion on the tank inductance	H
L_S	Source inductance (connected to the source of the transistor)	H
M	Mutual inductance	H
MDS	Minimum detectable signal of a receiver (also known as sensitivity)	dBm
n	Turn ratio of a transformer	
N	Number of stages in the receiver	

(continued)

Symbol	Description	Unit
$N_{Antenna}$	Available noise power coming from the antenna	W
$N_{i,tot}$	The total available noise power due to the receiver referred to its input	W
NF_{tot}	Noise factor of the whole receiver	dB
NPD	Noise plus distortion of a receiver referred to its input	W
P	Power consumption of a stage	W
P_{Ck}	Power coefficient of the kth stage: a proportionality constant relating the power consumption to the noise and linearity performance	W
$P_{IMDqi,tot}$	Equivalent input-referred available power of in-band q^{th} order intermodulation distortion due to an out-of-band interferer	W
P_S	Average power dissipated in the resistive part of the tank	W
P_{tot}	Power consumption of the whole receiver	W
Q	Effective quality factor of the tank	
Q_C	Quality factor of the tank capacitance	
Q_L	Quality factor of the tank inductance	
R_{CP}	Equivalent parallel resistance of the tank capacitance	Ω
R_{CS}	Series resistance of the tank capacitance	Ω
R_G	Gate resistance	Ω
R_{in}	Real part of the input impedance	Ω
R_L	Load resistance	Ω
R_{LP}	Equivalent parallel resistance of the tank inductance	Ω
R_{LS}	Series resistance of the tank inductance	Ω
r_o	Output resistance of the transistor	Ω
R_{out1}	The real part of the impedance seen from the positive node of the differential output	Ω
R_{out2}	The real part of the impedance seen from the negative node of the differential output	Ω
R_S	Source resistance	Ω
$S_x^{F(x)}$	Normalized single point sensitivity of F(x) to the variable x	
SNDR	Signal to noise-plus-distortion ratio	dB
T	Absolute temperature	K
T_0	Reference temperature for noise figure definition: 290°K	K
TR	Tuning range of a VCO	
U	Mason invariant: the gain of a linear two-port after unilateraliztion by embedding in a linear lossless reciprocal four-port	
V_{gs}	Gate-source voltage	V
$V_{imd2out}$	Differential output IMD2 voltage of the mixer	V
$V_{IMDqi,k}$	Equivalent input-referred voltage of the qth order intermodulation distortion of the kth stage of the receiver	V
$V_{IMDqi,tot}$	Equivalent input-referred voltage of the qth order intermodulation distortion of the whole receiver	V
$V_{interferer}$	Voltage of worst-case out-of-band interferer signal at the input of the receiver	V
$V_{IPqi,k}$	Voltage of qth order input intercept point of the kth stage of the receiver	V
$V_{IPqi,tot}$	Voltage of qth order input intercept point of the whole receiver	V
V_{LO}	Local oscillator voltage connected to the LO port of the mixer	V

(continued)

Symbol	Description	Unit
$V_{ni,k}$	Equivalent input-referred noise voltage of the kth stage of the receiver	$V/\sqrt{Hz}$
$V_{ni,tot}$	Equivalent input-referred noise voltage of the whole receiver	$V/\sqrt{Hz}$
$V_{nout,k}$	Noise voltage due to the kth stage at the output of the stage and at the signal band	$V/\sqrt{Hz}$
$V_{nout,k,image}$	Noise voltage due to the kth stage at the output of the stage and at the image band	$V/\sqrt{Hz}$
$V_{nout,Rs}$	Noise at the output of the mixer due to the source resistance	$V/\sqrt{Hz}$
$V_{nout,RL}$	Noise at the output of the mixer due to the load resistance	$V/\sqrt{Hz}$
$V_{nout,sw}$	Noise at the output of the mixer due to the switching pair	$V/\sqrt{Hz}$
$V_{nout,tot}$	Noise voltage due to the whole receiver at the output of the receiver and at the signal band	$V/\sqrt{Hz}$
$V_{nout,k,image}$	Noise voltage due to the whole receiver at the output of the receiver and at the image band	$V/\sqrt{Hz}$
V_{off}	Equivalent voltage offset representing the mismatch between the transistors in a differential pair	V
V_{out-}	Voltage of the negative node of the differential output	V
V_{out+}	Voltage of the positive node of the differential output	V
V_{RF}	RF voltage connected to the RF port of the mixer	V
V_T	Threshold voltage of a MOS transistor	V
W	Transistor width	m
X_{in}	Imaginary part of the input impedance	Ω
Z	Impedance matrix of a two-port	Ω
Z_{in}	Input impedance	Ω
Z_{L1}	The impedance seen from the positive node of the differential output	Ω
Z_{L2}	The impedance seen from the negative node of the differential output	Ω
Z_{src}	Source impedance	Ω
α_k	Contribution factor: the ratio between the noise (or distortion) contribution of the kth stage to that of its following stage	
β	Proportionality constant between the current in an MOS transistor and the square of the difference between gate-source voltage and threshold voltage	A/V^2
γ	Transistor channel noise coefficient	
ΔNPD	Variations of the noise plus distortion of a receiver referred to its input	W
$\Delta V_{IPqi,k}$	Variations of the voltage of qth order input intercept point of the kth stage of the receiver	V
$\Delta V_{ni,k}$	Variations of equivalent input-referred noise voltage of the kth stage of the receiver	$V/\sqrt{Hz}$
ΔV_T	Absolute mismatch between the threshold voltage of the transistors in a differential pair	V
ΔZ_{in}	Variations of input impedance	Ω
δZ_L	Relative mismatch in the differential load impedance	
$\Delta\beta$	Absolute mismatch between the β of the transistors in a differential pair	A/V^2
$\Delta\theta$		$1/V$

(continued)

Symbol	Description	Unit
	Absolute mismatch between the θ of the transistors in a differential pair	
$\Delta\omega$	Offset from the oscillation angular frequency of the VCO	rad/s
$\Delta\omega_{1/f^3}$	Angular frequency of the corner between $1/f^3$ and $1/f^2$ regions	rad/s
θ	A fitting factor taking into account the short channel effects of MOS transistors	1/V
μ_n	The mobility of an NMOS transistor	$m^2/V/s$
τ	Time constant at the common-source node of the switching pair of the mixer	s
$\Phi(t)$	Phase of the sinusoidal output of a VCO	rad
$\varphi_{IMDqi,k}$	The phase of the equivalent input-referred voltage of the qth order intermodulation distortion of the kth stage	rad
$\varphi_{IMDqi,tot}$	The phase of the equivalent input-referred voltage of the qth order intermodulation distortion of the whole receiver	rad
$\varphi_{IPqi,k}$	Phase of the voltage of the qth order input intercept point of the kth stage	V
$\varphi_{IPqi,tot}$	Phase of the voltage of the qth order input intercept point of the whole receiver	V
$\varphi_{vn,k}$	The phase of the voltage gain of the kth stage at the signal band including the loading effects	rad
$\varphi_{vn,tot}$	The phase of the voltage gain of the whole receiver at the signal band	rad
ω	Angular frequency	rad/s
ω_0	Oscillation angular frequency of a VCO	rad/s
ω_1	Angular frequency of the first tone in a multi-tone signal	rad/s
ω_2	Angular frequency of the second tone in a multi-tone signal	rad/s
ω_3	Angular frequency of the third tone in a multi-tone signal	rad/s
ω_{center}	Center (the middle of the maximum and minimum) angular frequency of oscillation of a VCO	rad/s
ω_{IF}	IF Angular frequency	rad/s
ω_{LO}	LO Angular frequency	rad/s
ω_{max}	Maximum angular frequency of oscillation of a VCO	rad/s
ω_{min}	Minimum angular frequency of oscillation of a VCO	rad/s

Bibliography

1. Ramakrishnan H, Shedabale S, Russell G, Yakovlev A (2008) Analysing the effect of process variation to reduce parametric yield loss. IEEE international conference on integrated circuit design and technology, June 2008, pp 171–176
2. Springer SK, Lee S, Lu N, Nowak EJ, Plouchart J-O, Watts JS, Williams RQ, Zamdmer N (2006) Modeling of variation in submicrometer CMOS ULSI technologies. IEEE Trans Electron Dev 53(9):2168–2178
3. Sery G, Borkar S, De V (2002) Life is CMOS: why chase the life after? Proceedings of 39th design automation conference, pp 78–83
4. Karnik T, Borkar S, De V (2002) Sub-90 nm technologies-challenges and opportunities for CAD. IEEE/ACM international conference on computer aided design, Nov 2002, pp 203–206
5. Borkar S, Karnik T, Narendra S, Tschanz J, Keshavarzi A, De V (2003) Parameter variations and impact on circuits and microarchitecture. Proceedings of design automation conference, June 2003, pp 338–342
6. International Technology Roadmap for Semiconductors (2011) http://public.itrs.net
7. Overclockers.com website, overclockers forum (2011) http://www.overclockers.com
8. Meehan MD, Purviance J (1993) Yield and reliability in microwave circuit and system design. Artech House, Boston
9. Strojwas AJ (2006) Conquering process variability: a key enabler for profitable manufacturing in advanced technology nodes. IEEE international symposium on semiconductor manufacturing, Sep. 2006, pp xxiii–xxxii
10. Pang L-T, Qian K, Spanos CJ, Nikolic B (2009) Measurement and analysis of variability in 45 nm strained-Si CMOS technology. IEEE J Solid State Circuits 44(8):2233–2243
11. Austin T, Bertacco V, Blaauw D, Mudge T (2005) Opportunities and challenges for better than worst-case design. Proceedings of the Asia and South Pacific design automation conference, Jan 2005, pp I/2–I/7
12. Sheng W, Emira A, Sanchez-Sinencio E (2006) CMOS RF receiver system design: a systematic approach. IEEE Trans Circuits Syst-I: Reg Pap 53(5):1023–1034
13. Baltus P (2004) Minimum power design of RF front ends. PhD Dissertation, Eindhoven University of Technology
14. El-Nozahi M, Sanchez-Sinencio E, Entesari K (2009) Power-aware multiband–multistandard CMOS receiver system-level budgeting. IEEE Trans Circuits Syst II: Exp Brief 56 (7):570–574
15. Part 15.3: wireless medium access control (MAC) and physical layer (PHY) specifications for high rate wireless personal area networks (WPANs): amendment 2: millimeter-wave based alternative physical layer extension. IEEE 802.15.3c, Oct 2009
16. Janssen E, Mahmoudi R, van der Heijden E, Sakian P, de Graauw A, Pijper R, van Roermund A (2010) Fully balanced 60 GHz LNA with 37% bandwidth, 3.8 dB NF, 10 dB gain and

constant group delay over 6 GHz bandwidth. 10th topical meeting on silicon monolithic integrated circuits in RF systems, Jan 2010

17. Sakian P, Mahmoudi R, van der Heijden E, de Graauw A, van Roermund A (2011) Wideband cancellation of second order intermodulation distortions in a 60 GHz zero-IF mixer. 11th topical meeting on silicon monolithic integrated circuits in RF systems, Jan 2011

18. Tomkins A, Aroca RA, Yamamoto T, Nicolson ST, Doi Y, Voinigescu SP (2009) A zero-IF 60 GHz 65 nm CMOS transceiver with direct BPSK modulation demonstrating up to 6 Gb/s data rate over a 2 m wireless link. IEEE J Solid State Circuits 44(8):2085–2099

19. Marcu C, Chowdhury D, Thakkar C, Park J-D, Kong L-K, Tabesh M, Wang Y, Afshar B, Gupta A, Arbabian A, Gambini S, Zamani R, Alon E, Niknejad AM (2009) A 90 nm CMOS low-power 60 GHz transceiver with integrated baseband circuitry. IEEE J Solid State Circuits 44(12):3434–3447

20. Alpman E, Lakdawala H, Carley LR, Soumyanath K (2009) A 1.1 V 50 mW 2.5 GS/s 7b time-interleaved C-2C SAR ADC in 45 nm LP digital CMOS. IEEE international solid state circuits conference, Feb 2009

21. Deng W, Mahmoudi R, Harpe P, van Roermund A (2008) An alternative design flow for receiver optimization through a trade-off between RF and ADC. IEEE radio wireless symposium, Jan 2008

22. Yang J, Lin Naing T, Brodersen RW (2010) A 1 GS/s 6 Bit 6.7 mW successive approximation ADC using asynchronous processing. IEEE J Solid State Circuits 45(8):1469–1478

23. Ashby KB, Koullias IA, Finley WC, Bastek JJ, Moinian S (1996) High Q inductors for wireless applications in a complementary silicon bipolar process. IEEE J Solid State Circuits 31(1):4–9

24. Yang R, Qian H, Li J, Xu Q, Hai C, Han Z (2006) SOI technology for radio-frequency integrated-circuit applications. IEEE Trans Electron Dev 53(6):1310–1316

25. Chang JY-C, Abidi AA, Gaitan M (1993) Large suspended inductors on silicon and their use in a 2-μm CMOS RF amplifier. IEEE Electron Dev Lett 14(5):246–248

26. Yue CP, Wong SS (1998) On-chip spiral inductors with patterned ground shields for Si-based RF ICs. IEEE J Solid-State Circuits 33(5):743–752

27. Yue CP, Ryu C, Lau J, Lee TH, Wong SS (1996) A physical model for planar spiral inductors on silicon. International electron devices meeting, Dec 1996, pp 155–158

28. Weisshaar A, Lan H, Luoh A (2002) Accurate closed-form expressions for the frequency-dependent line parameters of on-chip interconnects on lossy silicon substrate. IEEE Trans Adv Pack 25(2):288–296

29. Wiemer L, Jansen RH (1987) Determination of coupling capacitance of underpasses, air bridges and crossings in MICs and MMICs. Electron Lett 23(7):344–346

30. Cheung TSD, Long JR (2006) Shielded passive devices for silicon-based monolithic microwave and millimeter-wave integrated circuits. IEEE J Solid-State Circuits 41(5):1183–1200

31. Noise figure measurement accuracy- the Y-factor method (2010), Agilent application note 57-2. http://cp.literature.agilent.com/litweb/pdf/5952-3706E.pdf

32. Pozar DM (2005) Microwave engineering. Wiley, Hoboken

33. Tiemeijer LF, Pijper RMT, van der Heijden E (2010) Complete on-wafer noise-figure characterization of 60-GHz differential amplifiers. IEEE Trans Microw Theory Tech 58 (6):1599–1608

34. Agilent PSA series spectrum analyzers phase noise measurement personality (2005), technical overview with self-guided demonstration: option 226. http://cp.literature.agilent.com/litweb/pdf/5988-3698EN.pdf

35. Agilent's phase noise measurement solutions: finding the best fit for your test requirements, Selection Guide (2011). http://cp.literature.agilent.com/litweb/pdf/5990-5729EN.pdf

36. Agilent E5052A signal source analyzer: advanced phase noise and transient measurement techniques, application note (2004). http://cp.literature.agilent.com/litweb/pdf/5989-1617EN.pdf

37. Agilent E5052B signal source analyzer: advanced phase noise and transient measurement techniques, application note (2007). http://cp.literature.agilent.com/litweb/pdf/5989-7273EN.pdf
38. Walls WF (2001) Practical problems involving phase noise measurements. 33rd annual precise time and time interval meeting, Nov 2001, pp 407–416
39. Lee TH (2004) The design of CMOS radio-frequency integrated circuits. Cambridge University Press, Cambridge
40. Shaeffer DK, Lee TH (1997) A 1.5-V, 1.5-GHz CMOS low noise amplifier. IEEE J Solid-State Circuits 32(5):745–759
41. Yao T, Gordon MQ, Tang KKW, Yau KHK, Yang M-T, Schvan P, Voinigescu SP (2007) Algorithmic design of CMOS LNAs and PAs for 60-GHz radio. IEEE J Solid-State Circuits 42(5):1044–1057
42. Zhuo W, Embabi S, Pineda de Gyvez J, Sanchez-Sinencio E (2000) Using capacitive cross-coupling technique in RF low noise amplifiers and down-conversion mixer design. Proceedings of European solid state circuits conference, Sep. 2000, pp 116–119
43. Bruccoleri F, Klumperink EAM, Nauta B (2001) Generating all 2-MOS transistors amplifiers leads to new wide-band LNAs. IEEE J Solid-State Circuits 36(7):1032–1040
44. Bruccoleri F, Klumperink EAM, Nauta B (2004) Wide-band CMOS low-noise amplifier exploiting thermal noise canceling. IEEE J Solid-State Circuits 39(2):275–282
45. Cassan DJ, Long JR (2003) A 1-V transformer-feedback low-noise amplifier for 5-GHz wireless LAN in 0.18-μm CMOS. IEEE J Solid-State Circuits 38(3):427–435
46. Mason SJ (1954) Power gain in feedback amplifiers. Trans IRE Prof Group Circuit Theory CT-1(2):20–25
47. Gupta MS (1992) Power gain in feedback amplifiers, a classic revisited. IEEE Trans Microw Theory Tech 40(5):864–879
48. Singhakowinta A, Boothroyd AR (1964) On linear twoport amplifiers. IEEE Trans Circuit Theory CT-11(1):169
49. Cheema HM, Sakian P, Janssen E, Mahmoudi R, van Roermund A (2009) Monolithic transformers for high frequency bulk CMOS circuits. IEEE topical meeting on silicon monolithic integrated circuits in RF systems, Jan 2009
50. Cohen E, Ravid S, Ritter D (2008) An ultra low power LNA with 15 dB gain and 4.4 dB NF in 90 nm CMOS process for 60 GHz phase array radio. IEEE radio frequency integrated circuits symposium, June 2008, pp 61–64
51. Siligaris A, Mounet C, Reig B, Vincent P, Michel A (2008) CMOS SOI technology for WPAN. Application to 60 GHz LNA. IEEE international conference on integrated circuit design and technology and tutorial, June 2008
52. Borremans J, Raczkowski K, Wambacq P (2009) A digitally controlled compact 57-to-66 GHz front-end in 45 nm digital CMOS. IEEE international solid-state circuits conference – digest of technical papers, Feb 2009, pp 492–493
53. Weyers C, Mayr P, Kunze JW, Langmann U (2008) A 22.3 dB voltage gain 6.1 dB NF 60 GHz LNA in 65 nm CMOS with differential output. IEEE international solid-state circuits conference – digest of technical papers, Feb 2008, pp 192–606
54. Razavi B (1997) Design considerations for direct-conversion receivers. IEEE Trans Circuits Syst II: Analog Digit Signal Process 44(6):428–435
55. Gilbert B (1968) A precise four-quadrant multiplier with subnanosecond response. IEEE J Solid-State Circuits 3(4):365–373
56. Darabi H, Abidi AA (2000) Noise in RF-CMOS mixers: a simple physical model. IEEE J Solid-State Circuits 35(1):15–25
57. Lee S-G, Choi J-K (2000) Current-reuse bleeding mixer. Electron Lett 36(8):696–697
58. Vitali S, Franchi E, Gnudi A (2007) RF I/Q downconverter with gain/phase calibration. IEEE Trans Circuits Syst II: Exp Briefs 54(4):367–371
59. Razavi B (2006) A 60-GHz CMOS receiver front-end. IEEE J Solid-State Circuits 41 (1):17–22

60. Razavi B (1997) A 900-MHz CMOS direct conversion receiver. Symposium on VLSI circuits digest of technical papers, June 1997, pp 113–114
61. Heydari P (2004) High-frequency noise in RF active CMOS mixers. Proceedings of the Asia and South Pacific design automation conference, Jan 2004, pp 57–61
62. Cheng W, Annema AJ, Croon JA, Nauta B (2011) Noise and nonlinearity modeling of active mixers for fast and accurate estimation. IEEE Trans Circuits Syst I: Reg Pap 58(2):276–289
63. Weiner DD, Spina JE (1980) Sinusoidal analysis and modeling of weakly nonlinear circuits. Van Norstrand Reinhold, New York
64. Wambacq P, Sansen W (1998) Distortion analysis of analog integrated circuits. Kluwer, Boston
65. Terrovitis MT, Meyer RG (2000) Intermodulation distortion in current-commutating CMOS mixers. IEEE J Solid State Circuits 35(10):1461–1473
66. Manstretta D, Brandolini M, Svelto F (2003) Second-order intermodulation mechanisms in CMOS downconverters. IEEE J Solid State Circuits 38(3):394–406
67. Dufrene K, Boos Z, Weigel R (2007) A 0.13μm 1.5V CMOS I/Q downconverter with digital adaptive IIP2 calibration. IEEE international solid-state circuits conference digest of technical papers, Feb 2007, pp 86–589
68. Wang J, Wong AKK (2011) Effects of mismatch on CMOS double-balanced mixers: a theoretical analysis. 2001 IEEE Hong Kong electron devices meeting, Hong Kong, 2001
69. Zhang F, Skafidas E, Shieh W (2007) A 60-GHz double-balanced Gilbert cell down-conversion mixer on 130-nm CMOS. IEEE radio frequency integrated circuits symposium, Honolulu, June 2007
70. Parsa A, Razavi B (2009) A new transceiver architecture for the 60-GHz band. IEEE J Solid-State Circuits 44(3):751–762
71. Rofougaran A, Rael J, Rofougaran M, Abidi A (1996) A 900 MHz CMOS LC oscillator with quadrature outputs. IEEE international solid-state circuits conference digest of technical papers, Feb 1996, pp 392–393
72. Laskin E, Rylyakov A (2009) A 136-GHz dynamic divider in SiGe technology. IEEE topical meeting on silicon monolithic integrated circuits in RF systems, Jan 2009
73. Ng AWL, Luong HC (2007) A 1-V 17-GHz 5-mW CMOS quadrature VCO based on transformer coupling. IEEE J Solid-State Circuits 42(9):1933–1941
74. Cheema HM, Mahmoudi R, van Roermund A (2010) On the importance of chip-level EM-simulations for 60-GHz CMOS circuits. European microwave integrated circuits conference, Sep. 2010, pp 246–249
75. Cheema HM (2010) Flexible phase-locked loops and millimeter wave PLL components for 60-GHz wireless networks in CMOS. PhD Dissertation, Eindhoven University of Technology
76. Baghdady EJ, Lincoln RN, Nelin BD (1965) Short-term frequency stability: characterization, theory, and measurement. Proc IEEE 53(7):704–722
77. Cutler LS, Searle CL (1966) Some aspects of the theory and measurement of frequency fluctuations in frequency standards. Proc IEEE 54(2):136–154
78. Leeson DB (1966) A simple model of feedback oscillator noise spectrum. Proc IEEE 54(2):329–330
79. Rael JJ, Abidi AA (2000) Physical processes of phase noise in differential LC oscillators. Proceedings of the IEEE custom integrated circuits conference, 2000, pp 569–572
80. Hajimiri A, Lee TH (1998) A general theory of phase noise in electrical oscillators. IEEE J Solid-State Circuits 33(2):179–194
81. Hajimiri A, Lee TH (1998) Corrections to "A general theory of phase noise in electrical oscillators". IEEE J Solid-State Circuits 33(6):928
82. Andreani P, Xiaoyan W, Vandi L, Fard A (2005) A study of phase noise in colpitts and LC-tank CMOS oscillators. IEEE J Solid-State Circuits 40(5):1107–1118
83. Andreani P, Fard A (2006) More on the $1/f^2$ phase noise performance of CMOS differential-pair LC-tank oscillators. IEEE J Solid-State Circuits 41(12):2703–2712
84. Fard A, Andreani P (2007) An analysis of $1/f^2$ phase noise in bipolar colpitts oscillators (with a digression on bipolar differential-pair LC oscillators). IEEE J Solid-State Circuits 42(2):374–384

85. Mazzanti A, Andreani P (2008) Class-C harmonic CMOS VCOs, with a general result on phase noise. IEEE J Solid-State Circuits 43(12):2716–2729

86. Murphy D, Rael JJ, Abidi AA (2010) Phase noise in LC oscillators: a phasor-based analysis of a general result and of loaded Q. IEEE Trans Circuits Syst I: Reg Pap 57(6):1187–1203

87. Kim DD, Kim J, Plouchart JO, Cho C, Li W, Lim D, Trzcinski R, Kumar M, Norris C, Ahlgren D (2007) A 70 GHz manufacturable complementary LC-VCO with 6.14 GHz tuning range in 65 nm SOI CMOS. IEEE international solid-state circuits conference digest of technical papers, Feb 2007, pp 540–541

88. Huang D, Hant W, Wang N-Y, Ku TW, Gu Q, Wong R, Chang MC-F (2006) A 60 GHz CMOS VCO using on-chip resonator with embedded artificial dielectric for side, loss and noise reduction. IEEE international solid-state circuits conference digest of technical papers, Feb 2006, pp 314–315

89. Parvais B, Scheir K, Vidojkovic V, Vandebriel R, Vandersteen G, Soens C, Wambacq P (2010) A 40 nm LP CMOS PLL for high-speed mm-wave communication. 2010 Proceedings of the 36th European solid-state circuits conference, Sep. 2010

90. Bozzola S, Guermandi D, Mazzanti A, Svelto F (2008) An 11.5% frequency tuning, -184dBc/Hz noise FOM 54 GHz VCO. IEEE radio frequency integrated circuits symposium, June 2008, pp 657–660

91. Musa A, Murakami R, Sato T, Chiavipas W, Okada K, Matsuzawa A (2010) A 58-63.6GHz quadrature PLL frequency synthesizer in 65nm CMOS. 2010 IEEE Asian solid state circuits conference, Nov 2010

92. Notten MGM, Veenstra H (2008) 60GHz quadrature signal generation with a single phase VCO and polyphase filter in a 0.25µm SiGe BiCMOS technology. IEEE bipolar/BiCMOS circuits and technology meeting, Oct 2008, pp 178–181

93. Ellinger F, Morf T, Buren GV, Kromer C, Sialm G, Rodoni L, Schmatz M, Jackel H (2004) 60GHz VCO with wideband tuning range fabricated on VLSI SOI CMOS technology. IEEE MTT-S international microwave symposium digest of technical papers, vol 3, pp 1329–1332

94. Sivonen P, Vilander A, Parssinen A (2005) Cancellation of second-order intermodulation distortion and enhancement of IIP2 in common-source and common-emitter RF transconductors. IEEE Trans Circuits Syst I: Reg Pap 52(2):305–317

95. Bautista EE, Bastani B, Heck J (2000) A high IIP2 downconversion mixer using dynamic matching. IEEE J Solid-State Circuits 35(12):1934–1941

96. Chen M, Wu Y, Chang MF (2006) Active 2nd-order intermodulation calibration for direct-conversion receivers. IEEE international solid-state circuits conference digest of technical papers, Feb 2006, pp 1830–1839

97. Dufrene K, Weigel R (2006) A novel IP2 calibration method for low-voltage downconversion mixers. IEEE radio frequency integrated circuits symposium, June 2006

98. Kivekas K, Parssinen A, Ryynanen J, Jussila J, Halonen K (2002) Calibration techniques of active BiCMOS mixers. IEEE J Solid State Circuits 37(6):766–769

99. Hotti M, Ryynanen J, Kievekas K, Halonen K (2004) An IIP2 calibration technique for direct conversion receivers. IEEE international symposium on circuits and systems, May 2004

100. Ler C-L, bin A'ain AK, Kordesch AV (2008) Compact, high-Q, and low-current dissipation CMOS differential active inductor. IEEE Microw Wirel Compon Lett 18(10):683–685

101. Verbruggen B, Craninckx J, Kuijk M, Wambacq P, Van der Plas G (2010) A 2.6mW 6b 2.2GS/s 4-times interleaved fully dynamic pipelined ADC in 40nm digital CMOS. IEEE international solid-state circuits conference digest of technical papers, Feb 2010, pp 296–298

102. Lont M, Mahmoudi R, van der Heijden E, de Graauw A, Sakian P, Baltus P, van Roermund A (2009) A 60GHz Miller effect based VCO in 65nm CMOS with 10.5% tuning range. IEEE topical meeting on silicon monolithic integrated circuits in RF systems, 19–21 Jan 2009

103. Stadius K, Kaumisto R, Porra V (1999) Varactor diodeless harmonic VCOs for GHz-range applications. ICECS 1:505–508